Lecture Notes in Nanoscale Science and Technology

Volume 10

For further volumes:
http://www.springer.com/series/7544

Lecture Notes in Nanoscale Science and Technology

Volume 10

Series Editors:

Zhiming M. Wang
State Key Laboratory of Electronic Thin Film and Integrated Devices,
University of Electronic Science and Technology,
Chengdu, People's Republic of China

Andreas Waag
Institut für Halbleitertechnik, TU Braunschweig, Braunschweig, Germany

Gregory Salamo
Department of Physics, University of Arkansas, Fayetteville, AR, USA

Naoki Kishimoto
Quantum Beam Center, National Institute for Materials Science, Tsukuba,
Ibaraki, Japan

Stefano Bellucci
Laboratori Nazionali di Frascati, Istituto Nazionale di Fisica Nucleare, Frascati,
Italy

Young June Park
School of Electrical Engineering, Seoul National University, Shinlim Dong,
Kwanak-Gu, Seoul, Korea

For further volumes:
http://www.springer.com/series/7544

Aimé Lay-Ekuakille

Optical Waveguiding and Applied Photonics

Technological Aspects, Experimental Issue Approaches and Measurements

Springer

Aimé Lay-Ekuakille
Dipartimento d'Ingegneria dell'Innovazio
University of Salento
Lecce, Italy

ISSN 2195-2159 ISSN 2195-2167 (electronic)
ISBN 978-1-4614-5958-3 ISBN 978-1-4614-5959-0 (eBook)
DOI 10.1007/978-1-4614-5959-0
Springer New York Heidelberg Dordrecht London

Library of Congress Control Number: 2012951429

Printed on acid-free paper

Springer is part of Springer Science+Business Media (www.springer.com)

I thank Alessandro Massaro for his valuable help and advice for making this book a reality.

Contents

Chapter 1
Introduction

Exploration of the world of micro-/nanotechnology, thanks to recent developments, has given an added value to optoelectronics. The main goal of this book is to provide tools and methods for applied research in this specific field. All topics will be connected with each other in the book in order to guide the researcher in the creation of a research platform based on available facilities. A reliable approach to creating an operating platform of research is to combine technological know-how, design, and experimental aspects. All fields communicate by feedback systems. Design represents the first step and starts after a preliminary study of the given device considering the available technology and the final application. Fabrication will introduce new techniques on the basis of innovative materials, and technological optimization will guarantee a standard procedure regarding the final production of the device. Finally, experimentation on and accurate analysis of the measured outputs will round out quality control of the device. Feedback systems are applied to implement devices with well-defined input/output characteristics. The book will provide approaches to the implementation of stable optoelectronic devices where the stability can be verified by the simultaneous agreement of experimental, theoretical, and numerical results. In particular, theoretical tools will define the application and a proper layout of the device, in addition to the numerical results, will provide solutions including the prediction of fabrication errors (analysis of error margins) such as geometrical resolutions and layer thicknesses. Finally, experimental setups and signal processing will complete the research activity. The proposed research platform procedures can be applied to devices such as electromagnetic waveguides, photonic crystals, microelectromechanical systems (MEMS), optoelectronic systems, and applied electronics.

The list of "connected" topics studied in the book is as follows:

1. Basic principles of electromagnetism useful for sensing
2. Optical and microwave theories/approaches (transmission line theory, design approaches)
3. Theory and design of optoelectronic waveguides oriented toward technology

A. Lay-Ekuakille, *Optical Waveguiding and Applied Photonics*, Lecture Notes in Nanoscale Science and Technology 10, DOI 10.1007/978-1-4614-5959-0_1,

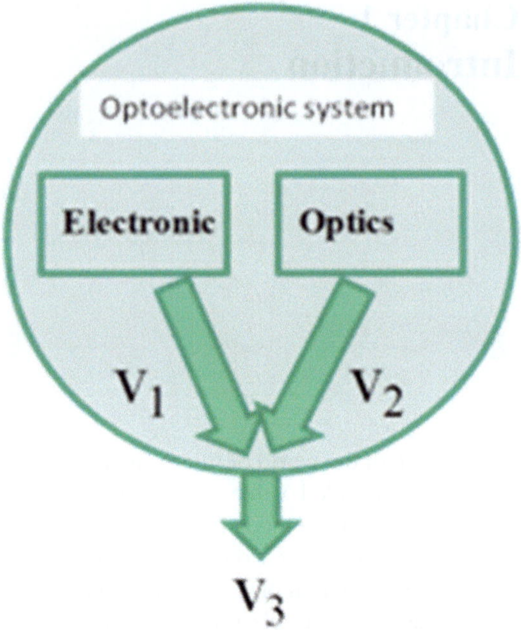

Fig. 1.1 Optoelectronic system and speed device response

4. Applications: sensors, couplers, fiber optics, integrated optics, discrete optics, birefringence, optical reflectors, biomedical devices, light trapping, distributed Bragg reflectors (DBRs), plasmonic antennas, wireless sensing, nonlinear optics, photonic crystals, nanoantennas, optical gratings, and cavities for light enhancement
5. Technology: nanotechnology, innovative nanocomposite materials, and fabrication machines
6. Innovative analytical/numerical approaches to the analysis of metallic/dielectric discontinuities in optical waveguides
7. Aspects of modeling and design in accordance with experimental setups
8. Applied electronics

An important aspect highlighted in the book is the matching of electronic components with optical devices characterized by different time responses, as indicated in Fig. 1.1.

Chapter 2
Optical Theory and Introduction to Optoelectronics

Abstract Theory is the "soul" of design: theory represents a basic understanding that predicts the real behavior of optical/optoelectronic devices and describes the physical aspects observed in experimental results. Starting from theoretical results, it is possible to optimize a designed device by means of numerical methods. A complete match between analytical, numerical, and experimental results will guarantee a correct operation of the analyzed system. Regarding guiding aspects of the waveguide at optical frequencies, in this chapter we will present a theoretical overview that includes geometrical optics, modal approaches, analytical approximations, coupled mode theory, quantum optics, photonic crystal mode theory, equation modeling, and scattering wave analysis. In particular, we will focus on theoretical approaches explaining the physical aspects and on some basic optical structures. An introduction to optoelectronic components will conclude the chapter. This chapter will help readers apply theory and know what the basic optoelectronic devices are. Optical devices and circuits are analyzed. Finally, an overview of important numerical methods and the main modeling aspects are provided.

2.1 Basic Concepts of Geometrical and Guided Optics

To introduce guided modes of dielectric waveguides (an example of an optical waveguide is shown in Fig. 2.1a illustrating a symmetrical slab waveguide), we first analyze, from a ray optics point of view, the interface between two different materials of refractive indices n_1 and n_2. A ray impinging from a core region (region with refractive index n_1) on the interface between two media is transmitted in the cladding (region with refractive index n_2) at an angle θ_2 with the following Snell's law (Rozzi and Mongiardo 1997):

$$n_1 \sin\theta_a = n_2 \sin\theta_b. \tag{2.1}$$

A. Lay-Ekuakille, *Optical Waveguiding and Applied Photonics*, Lecture Notes in Nanoscale Science and Technology 10, DOI 10.1007/978-1-4614-5959-0_2,

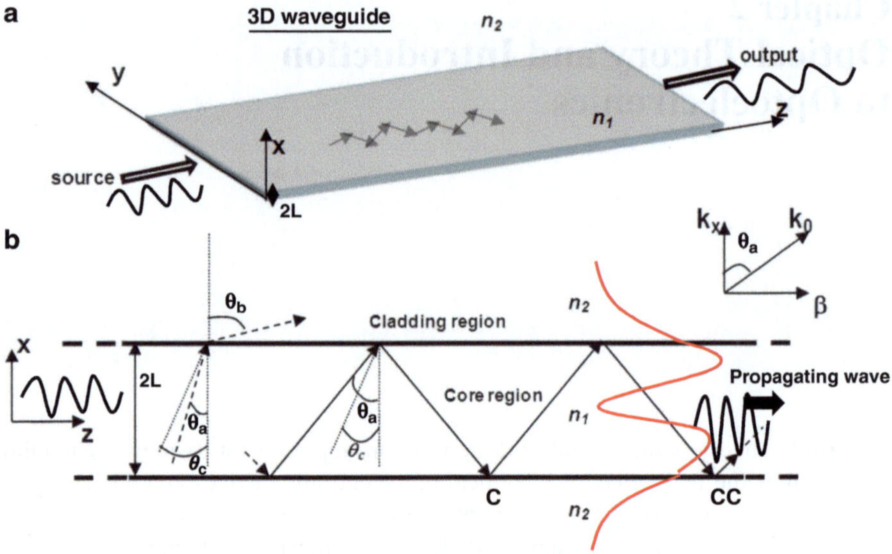

Fig. 2.1 (**a**) 3D scheme of symmetrical slab optical waveguide. An optical ray propagates in the core region (refractive index n_1) delimited by the cladding regions (refractive index n_2), which in the symmetrical waveguides are equal. (**b**) Snell's law angles defining refraction and guiding conditions

This ray is generated by an external optical source that can be a focused laser beam, a led optical source, or a generic lamp working at different optical frequency ranges. If medium 1 is denser than medium 2 ($n_1 > n_2$), then $\theta_b > \theta_a$ and there will be an angle of incidence (critical angle) $\theta_a = \theta_c = \arcsin(n_2/n_1)$ for a ray coming from 1 such that $\theta_b = \pi/2$. A ray impinging from the core region at an angle $\theta_a < \theta_c$ will appear in region 2 as refracted ray (Fig. 2.1a), and a ray impinging at an angle $\theta_a > \theta_c$ will undergo total reflection. If medium 1 is confined as illustrated in Fig. 2.1a (like a slab with core thickness equal to $2L$) and $\theta_a > \theta_c$, then the ray will be guided along medium 1 as a propagating mode: the ray undergoing total reflection at the upper interface will hit the lower interface again at the same angle (like a ray propagated in parallel metallic plate waveguide). Assuming that the transverse distribution of the propagating mode is the same at points C and CC, we obtain the condition for discrete values of incidence angles θ_i of guided modes,

$$\theta_i = \arccos \frac{n\pi}{2k_0L}, \tag{2.2}$$

obtained by assuming that the phase of a round trip in the transverse x-direction is a multiple of 2π, where k_0 is the wavevector along the direction of the ray (Fig. 2.1b). Equation (2.2) indicates that the guided modes of a slab waveguide can be described by a discrete set of independent rays, each one propagating along the longitudinal z-direction. The longitudinal direction is defined as the direction where the major

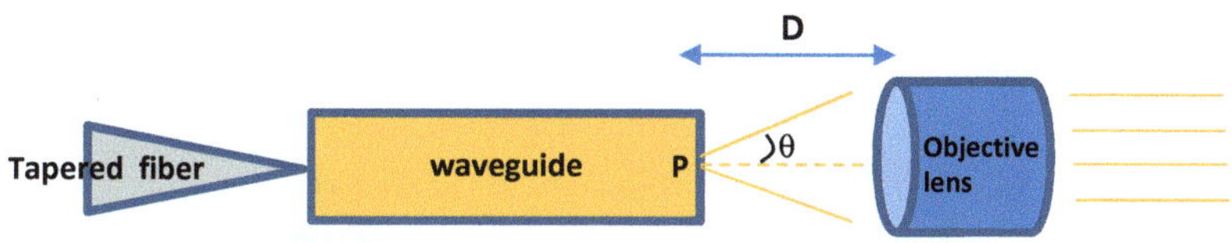

Fig. 2.2 Scheme of experimental setup measuring transmittivity of an optical integrated waveguide. Inset: angle θ defining numerical aperture

propagating electromagnetic power is directed. Other parts of powers are lost inside the cladding region or radiated outside as radiation modes. Each mode (guided, or radiation mode) will be identified by propagation constants in each direction defined by the *xyz* global coordinate system.

Each mode propagating in a slab waveguide can be identified by the following modal propagation constant in the *x*-direction:

$$k_x^2 = \omega^2 \mu \varepsilon + \gamma_z^2, \tag{2.3}$$

where μ is the magnetic permeability (H/m), ε the dielectric permittivity (F/m), ω the angular frequency, and γ_z the propagation constant of the guide in the *z*-direction. The slab waveguide supports transverse electric (TE) and transverse magnetic (TM) modes, where TE modes are characterized by ($\mathbf{H_x}$, $\mathbf{H_x}$, $\mathbf{E_y}$) and TM modes by ($\mathbf{E_x}$, $\mathbf{E_x}$, $\mathbf{H_y}$). The modes will transfer the optical power of the input source to the output of the waveguides and characterize the optical signals, which will propagate faster than the electric ones. This aspect highlights the different time responses between optical and electronic circuits: a good optoelectronic component optimizes the gap between the different time responses by means of a proper technology, including micro and miniaturized components able to match the micro-scale core region with the millimeter scale of the standard electronic components.

A parameter characterizing the source and the output power is the numerical aperture (NA). In optics, the NA of an optical system is a dimensionless number that characterizes the range of angles over which the system can accept or emit light. In most areas of optics, and especially in microscopy, the numerical aperture of an optical system, such as an objective lens, is defined by

$$\mathrm{NA} = n_i \sin\theta, \tag{2.4}$$

where n_i is the refraction index of the medium in which the system is working, and θ is the half-angle of the maximum cone of light. Concerning the measurement of the slab waveguide in Fig. 2.1a, the NA aperture characterizes the light coupling of the input and the output of the waveguide. For the input it is preferable to use a fiber with a tapered profile; in addition, in accordance with the characteristics of the output objective lens, the distance *D* must be fixed in order to improve a high collected power. The single-mode fibers should be used at the input in order to excite the waveguide by a micrometer spot size (Fig. 2.2).

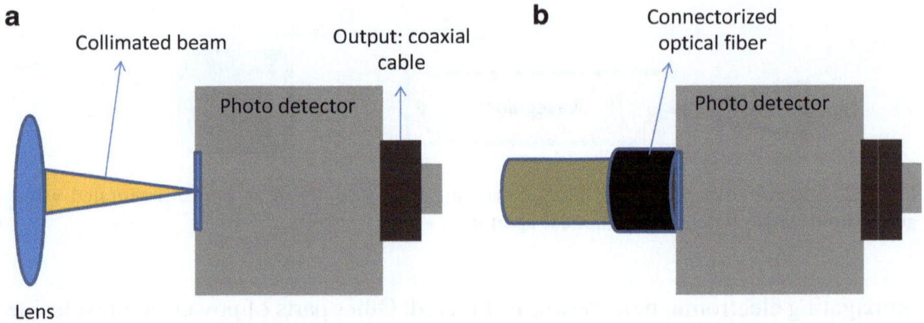

Fig. 2.3 (**a**) Photodetector with sensitive area coupled to collimated optical beam. (**b**) Photodetector with sensitive area coupled to connectorized optical fiber

A complete transmission light experimental setup for an optical waveguide could be described as follows: a light probe beam is launched from a tapered fiber and directly injected into the waveguide core. The light exiting the sample is collected and collimated by a microscope objective with high NA. An optical system made by two lenses with a horizontal slit could facilitate the light beam collimation on an optical waveguide, where the core thickness is on the order of the micro/nano scale. This collimation makes it possible to separate the light coming from the ridge waveguide from the radiation freely propagating in the air and through the substrate. At the output, the transmitted light could be collected by a multimode fiber with its free end lying on the focal plane of a lens (end-fire coupling) and brought to an optical analyzer based on optical intensity counting. By substituting the optical multimode fiber and the optical analyzer with a proper camera, it is possible to observe the mode profiles characterized by different spatial light distributions (a single light spot represents the fundamental mode, and multiple spot configurations indicate higher-order modes). With a photodetector as the analyzer, the optical ray could be coupled to the input by means of a lens or by an optical fiber connectorized to the sensitive area by a holder (Fig. 2.3). The optical signal will be converted into an electrical one and transmitted by a coaxial cable to an oscilloscope able to read the voltage intensity. The whole optical system (lenses, photo detector, optical fibers, etc.) is selected by the working wavelength band of the optical source: all the optical components are characterized by a wavelength band, which should be contained as illustrated in Fig. 2.4a, b. The transmitted optical output is mainly defined by the Boolean intersection between all the optical bands of all the components if there is a component with an optical response "out" the other optical bands, no signal will be transmitted (see the case of Fig. 2.5, where the lens is out from a Boolean intersection). The power of the transmitted will include all the guided modes or plane waves that are contained in the transmitted band and able to carry an optical power.

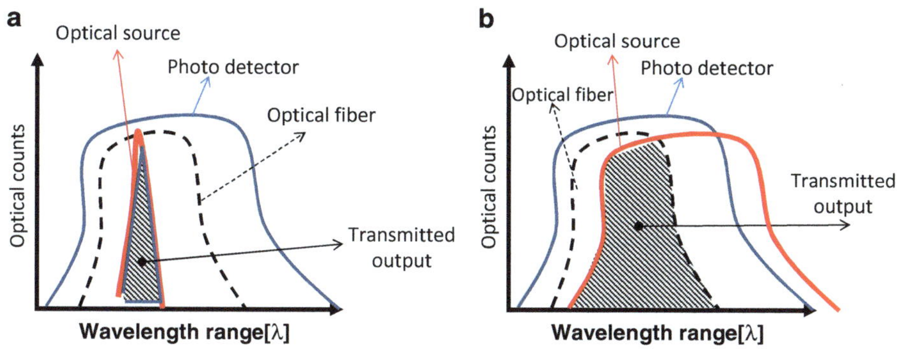

Fig. 2.4 (**a**) Optical responses and transmitted output coinciding with optical source band. (**b**) Optical responses and transmitted output partially coinciding with optical source

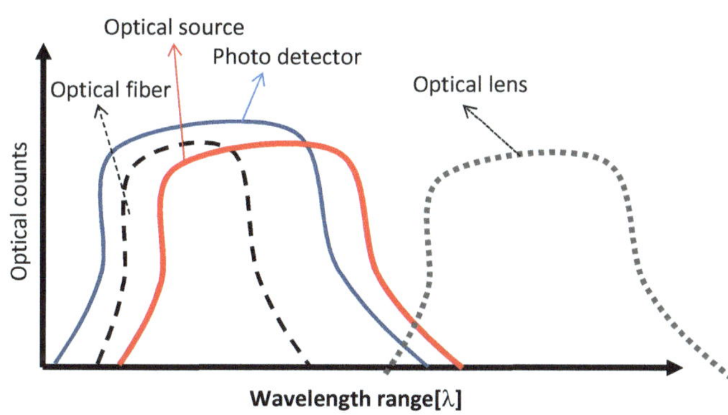

Fig. 2.5 Case where no transmitted signal is observed

2.2 Optical Modes and Measurement Approaches

2.2.1 Modes of a Slab Waveguide

Optical modes can be propagated into a waveguide that will be coupled to an optical system. In this case, the Boolean intersection with the waveguide optical response must also be considered. In this direction, the design of the waveguide is necessary in order to know the working frequencies associated with the guided modes able to transmit a signal. To understand how the modes carry electromagnetic energy, we define in this section the modes of a slab waveguide. Starting with the slab waveguide of Fig. 2.4 (uniform in the y-direction), we define two independent sets of three component TE and TM fields.

The set of equations is obtained by the following Maxwell curl equations (Rozzi and Mongiardo 1997):

$$\begin{aligned}\nabla \times \mathbf{E} &= -j\omega\mu\mathbf{H}\\ \nabla \times \mathbf{H} &= j\omega\varepsilon\mathbf{E}\end{aligned} \tag{2.5}$$

By assuming a traveling wave along the z-direction, the equivalence $\mathbf{E} = \mathbf{V}$ (voltage vector signal), and $\mathbf{H} = \mathbf{I}$ (current vector signal), the TE modes are defined by the following system of equations:

$$\begin{aligned}&k_z V_y = -j\omega\mu I_x,\\ &\frac{dV_y}{dx} = -j\omega\mu I_z,\\ &\frac{dI_z}{dx} = \left(\frac{k_z^2}{j\omega\mu} - j\omega\varepsilon\right)V_y,\end{aligned} \tag{2.6}$$

while the TM modes are defined by

$$\begin{aligned}&k_z I_y = j\omega\varepsilon V_x,\\ &\frac{dI_y}{dx} = j\omega\varepsilon V_z,\\ &\frac{dV_z}{dx} = \left(-\frac{k_z^2}{j\omega\varepsilon} + j\omega\mu\right)I_y,\end{aligned} \tag{2.7}$$

where $k_z = \alpha_z + j\beta_z$ is the propagation constant of the guide in the z-direction.

We observe that the analogy $\mathbf{E} = \mathbf{V}$ and $\mathbf{H} = \mathbf{I}$ is possible because, by neglecting the radiation modes of a discontinuous waveguide (core/air or cladding interface), we can express the global transverse field at any longitudinal position $z = z_0$ as a superposition of the discrete components (local eigenmodes) in two polarizations (Rozzi and Farina 1999),

$$\mathbf{E} = \sum_m V_m(z)\mathbf{E}_m(x, y), \tag{2.8}$$

$$\mathbf{H} = \sum_m I_m(z)\mathbf{H}_m(x, y), \tag{2.9}$$

normalized as follows:

$$\int_A \overline{E_c} \times \overline{H_p} \cdot \mathbf{z}\, ds = \begin{matrix} 1 \text{ if } c = p, \\ 0 \text{ if } c \neq p, \end{matrix} \tag{2.10}$$

where the integration is carried out over the cross sections A of the guide. Equations (2.9) and (2.10) could represent the guided modes as independent voltage and

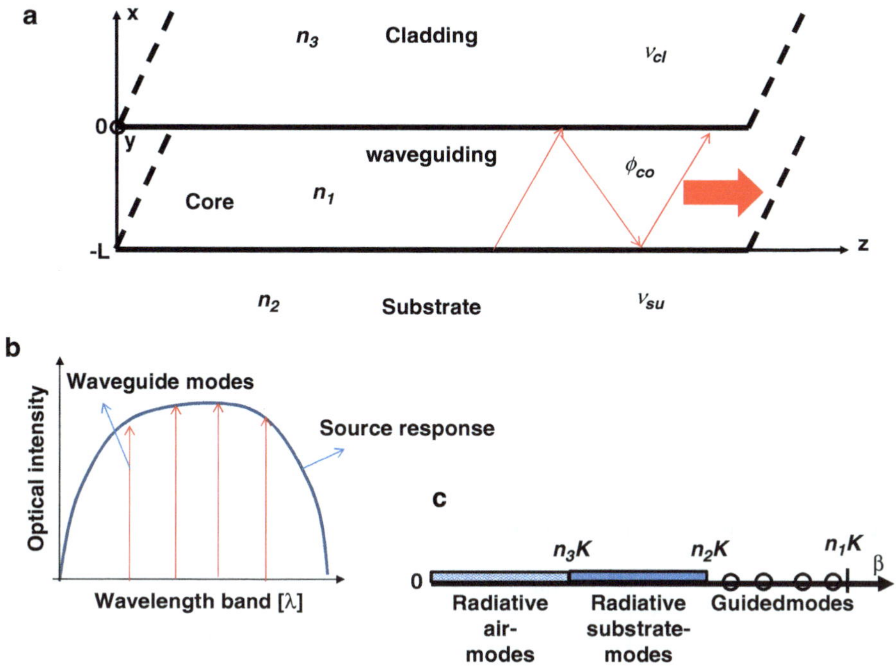

Fig. 2.6 (**a**) Scheme of asymmetrical slab waveguide. (**b**) Guided modes as wavelengths contained in source optical response. (**c**) Mode classification of asymmetrical slab waveguides

current signals traveling in decoupled transmission lines, where each transmission line represents a guided mode.

For the asymmetric slab waveguide of Fig. 2.6 with $n_1 > n_2 > n_3$, the guided TE mode, the V_y component, the solution of (2.7), can be written as (Marcuse 1974)

$$V_y(x) \propto \begin{cases} V_{cl}\exp\{-\nu^{TE}{}_{cl}x\} & (x > 0), \\ V_{co}\exp\{\gamma^{TE}{}_{co}x + \phi^{TE}{}_{co}\} & (-L < x < 0), \\ V_{su}\exp\{+\nu^{TE}{}_{su}(x+L)\} & (x < -L), \end{cases} \tag{2.11}$$

where V_{cl}, V_{co}, and V_{su} are constants, and $\nu^{\mathrm{TE}}{}_{cl}$, $\gamma^{\mathrm{TE}}{}_{co}$, and $\nu^{\mathrm{TE}}{}_{su}$ are the propagation constants of the cladding, core, and substrate region, respectively.

Regarding the guided TM mode, the I_y component, the solution of (2.8), can be written as

$$I_y(x) \propto \begin{cases} I_{cl}\exp\{-v_{cl}x\} & (x > 0), \\ I_{co}\exp\{\gamma_{co}x + \phi_{co}\} & (-L < x < 0), \\ I_{su}\exp\{+v_{su}(x+L)\} & (x < -L), \end{cases} \tag{2.12}$$

where I_{cl}, I_{co}, and I_{su} are constants, and $\nu^{\mathrm{TM}}{}_{cl}$, $\gamma^{\mathrm{TM}}{}_{co}$, and $\nu^{\mathrm{TM}}{}_{su}$ are the propagation constants of the cladding, core, and substrate region, respectively.

The symmetrical slab waveguide is a particular case of the asymmetrical slab waveguide. This case is characterized by $n_3 = n_2$ and consecutively by $\nu_{cl} = \nu_{su}$. By considering a core thickness of $2h$ and the origin of the coordinate system placed in the middle of the core, we obtain for the even TE mode the following characteristic equation:

$$\gamma_{co} \tan \gamma_{co} h = \nu_{su}, \tag{2.13}$$

and the distribution of V_y is defined by

$$V_y(x) \propto \begin{cases} \cos(\gamma^{TE}{}_{co} x) & \text{core region,} \\ \cos(\gamma^{TE}{}_{co} h) e^{-\nu^{TE}{}_{su}(x-h)} & \text{substrate/cladding.} \end{cases} \tag{2.14}$$

The odd TE modes are characterized by the following dispersion relation:

$$-\gamma^{TE}{}_{co} \tan \gamma^{TE}{}_{co} h = \nu_{su}, \tag{2.15}$$

and by the V_y distribution is defined by

$$V_y(x) \propto \begin{cases} \sin(\gamma^{TE}{}_{co} x) & \text{core region,} \\ \sin(\gamma^{TE}{}_{co} h) e^{-\nu_{su}(x-h)} & \text{substrate/cladding.} \end{cases} \tag{2.16}$$

Moreover, the dispersion relations for the TM-even and TM-odd modes are

$$\gamma^{TM}{}_{co} \tan \gamma^{TM}{}_{co} h = \nu_{su} \frac{n_1^2}{n_2^2}, \tag{2.17}$$

$$k_c \cot k_c d = -\gamma_s \frac{n_c^2}{n_s^2}, \tag{2.18}$$

respectively, and the TM-even (symmetric) and TM-odd (antisymmetric) mode distribution of I_y are the same as in (2.14) and (2.16).

2.2.2 *Modes of a Ridge Waveguide*

A ridge optical waveguide is commonly used in integrated optics in order to focus light in a region (ridge) that can transfer power efficiently. Others use this kind of waveguide to create semiconductor optical amplifiers, to integrate polymers behaving as Y-branch power splitters (especially for wavelength division multiplexing implementation in telecommunications systems), or to fabricate optoelectronic components such as ridge waveguide lasers.

In this section, we will describe a method to study the modes of a 3D waveguide. If the ridge is characterized by a periodic profile (Fig. 2.7a), then the waveguide

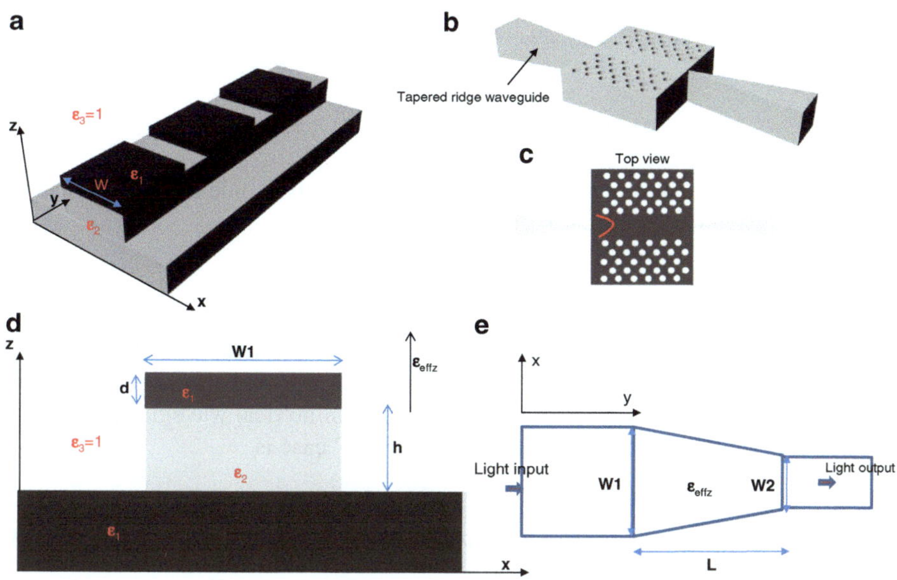

Fig. 2.7 (**a**) Ridge periodic waveguide. CAD (**b**) 3D view and (**c**) *top view* of ridge tapered waveguides exciting a photonic crystal waveguide. (**d**) Example of cross section of tapered waveguide. (**e**) Tapered layout model

behaves as an optical filter, a diffraction grating for light enhancement, or a photonic bandgap (PBG) device. If the waveguide profile is tapered, then the ridge waveguide behaves as an integrated light coupler able to transfer the maximum power in a guiding region or in another waveguide. An example of the application of the tapered waveguide is light coupling in line-defect photonic crystals (Fig. 2.7b, c).

To define the TE and TM modes of a ridge waveguide, we begin our analysis in the transverse z-direction in order to define the effective refractive index n_{effz} indicated in the layout of Fig. 2.7d. The dispersion equations could be used for a graphical analysis of the single-mode condition. This approach is useful for evaluating the sensitivity of the solution near the cutoff frequency (of the single TE guided condition) by changing the core thickness d. The dispersion equations we will use are (Marcuse 1974) as follows:

$$v_1(t) = \frac{\left(t^2 \tan(2t) - u\sqrt{{g_1}^2 - t^2}\right)}{t + \tan(2t) \cdot \sqrt{{g_1}^2 - t^2}} \tag{2.19}$$

for the TE modes and

$$v_1(t) = \frac{\left(t^2 \varepsilon_1 \varepsilon_3 \tan(2t) - t\varepsilon_2 \varepsilon_3 \sqrt{{g_1}^2 - t^2}\right)}{t\varepsilon_1 \varepsilon_3 + \varepsilon_2^2 \tan(2t) \cdot \sqrt{{g_1}^2 - t^2}} \tag{2.20}$$

for the TM modes, with

$$\begin{aligned} g_1 &= k_0 d\sqrt{\varepsilon_2 - \varepsilon_1}, \\ g_3 &= k_0 d\sqrt{\varepsilon_2 - \varepsilon_3}, \\ v_2(t) &= \sqrt{g_3^2 - t^2}, \\ k_0 &= \omega\sqrt{\varepsilon_0 \mu_0}. \end{aligned} \tag{2.21}$$

First to be evaluated is $k_z = t/d$ and then the effective index in the z-direction n_{effz} by the wave number conservation equation

$$k_0^2 \varepsilon_{effz} = k_0^2 \varepsilon_2 - k_z^2. \tag{2.22}$$

Further, the transverse propagation constant k_x is calculated by considering the symmetrical waveguide width W as the core thickness and ε_{effz} as the effective index. The dispersion equation of the equivalent symmetrical waveguide (slab with dielectric permittivity ε_{effz} in air) in the guided TE^x case is

$$\begin{aligned} t' \tan(t') &= p_1, \\ t'^2 + p_2{}^2 &= g'^2_1, \end{aligned} \tag{2.23}$$

and in the guided TM^x case it is

$$\begin{aligned} t' \tan(t') &= \frac{\varepsilon_{effz}}{\varepsilon_2} p_1, \\ t'^2 + p_2{}^2 &= g'^2_1, \end{aligned} \tag{2.24}$$

with

$$\begin{aligned} t' &= k_x W, \\ g_1' &= \sqrt{(\varepsilon_{effz} - \varepsilon_2)} k_0 W, \\ p_1 &= \nu_x W, \end{aligned} \tag{2.25}$$

and $\varepsilon_3 = \varepsilon_2 = 1$ (symmetrical slab), W is the slab thickness in the x-direction, and ν_x is the propagation constant in the air region. The ridge modal propagation constant in the y-propagated direction will be

$$\beta_y^2 = \varepsilon_1 k_0^2 - k_z^2 + k_x^2. \tag{2.26}$$

Figure 2.6 shows graphical approaches (Rozzi and Mongiardo 1997) to evaluating the transverse propagation constant in the z- and x-directions. The z-analysis permits us to evaluate, in the single TE^z mode condition, the effective index in order to analyze the 3D waveguide as a 2D structure. We show in Fig. 2.8 a theoretical graphic defining the effective refractive indices along the z-direction for different thicknesses d: the effective refractive index will increase with increases in parameter d. We observe that by increasing the thickness W at a fixed working wavelength λ_0 μm we can obtain other higher-order modes. Finally, we note that the geometrical

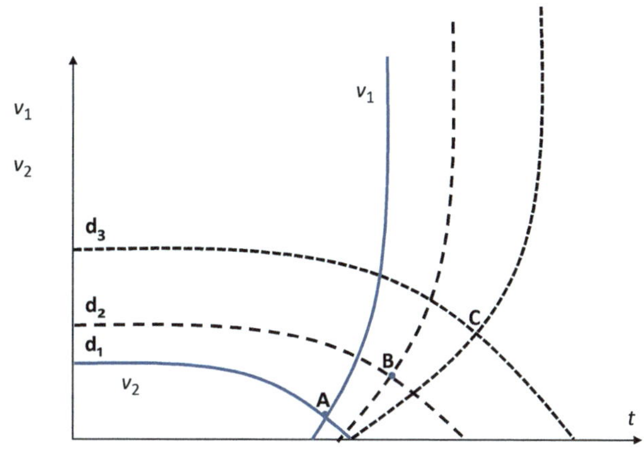

Fig. 2.8 Example of a theoretical graphical approach defining effective refractive indices for asymmetrical slab waveguide. In this case, $d_1 < d_2 < d_3$, and points *A*, *B*, and *C* define the three values of the effective refractive index n_{effz}. In the theoretical example, $n_{effz,A} < n_{effz,B} < n_{effz,C}$

parameter *L* of Fig. 2.7e changes the optical transmittivity of the tapered ridge waveguide defining the light coupling efficiency. Ridge waveguides are also used in optoelectronic integrating electrodes. The electrodes introduce an electro-optic effect: the voltage signal, applied on the ridge waveguide by means of electrodes, generates a change in the optical material property. This change consists of a variation in absorption or of a variation of the refractive indices. As for the variation in the refractive index, the voltage signal changes the solution of the effective refractive index and, consecutively, the mode propagating inside the waveguide (important aspect observed in the integrated polarizers). The main applications of the electro-optic effect are in the field of polarizers, amplitude modulators, Mach–Zehnder interferometers, and electro-optic deflectors. Other applications are in lasers: in recent research, electrodes are applied to photonic crystal lasers (made by quantum dots or using quantum well technology) supporting the output power.

2.2.3 Modes Coupled by Gratings

Gratings are commonly used in passive and active integrated optical circuits. The main applications are in Bragg gratings for filtering and laser applications. The grating design must consider the technology applied to the materials. Current technology allows the implementation of gratings also in optical fibers. If nonlinear materials are analyzed (as in second harmonic nonlinear processes), the grating should satisfy some conditions such as the quasi-phase matching (QPM) able to supply energy a second harmonic signal generated by the $\chi^{(2)}$ tensor of the material (Venugopal et al. 2004). An example of nonlinear grating where a mode coupling between a fundamental mode and a second harmonic is observed is the periodically switched nonlinearity (PSN) (Artigas et al. 2004) related to a $GaAs/Al_{0.4}Ga_{0.6}As$ structure. The grating is used also

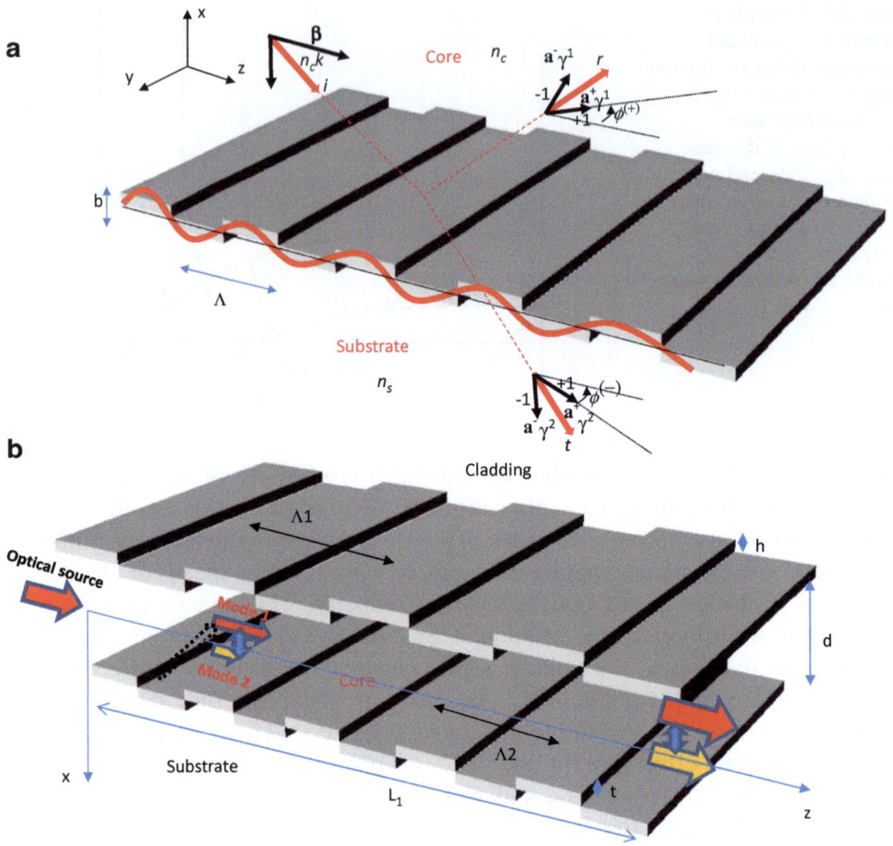

Fig. 2.9 (**a**) Scattering at core/substrate interface. The sinusoidally distorted core/substrate interface acts like a phase grating. (**b**) Double grating modeling of a symmetrical waveguide and coupling between two guided modes

for optical deflectors able to generate scattered plane waves, as illustrated in Fig. 2.9a, and for multimode coupling process by multiple grating (Fig. 2.9b).

If the grating is between the core and the cladding region or between the core and the substrate region, a proper phase-matched grating could couple efficiently a guided mode with a radiation mode. The following coupling coefficient $k_{G,R}$ will define the power exchanged between the two modes:

$$k_{G,R} = \int V_{y,Guided}(x)\Delta\varepsilon V_{y,Radiated}(x)dx, \tag{2.27}$$

where Δε represents the dielectric permittivity variation, which can be expanded by Fourier series (see the approximation of the sinusoidally core/substrate interface of Fig. 2.9b). Considering a 3D grating the coupling of energy between two modes, V_a

and V_b, will be characterized along the z-propagation direction by the following coupling coefficient:

$$k_{a,b}(z) = \omega \int_{-\infty}^{+\infty} \int_{-\infty}^{+\infty} \Delta\varepsilon(z) V_a \cdot V^*{}_b dxdy \tag{2.28}$$

2.3 Optical Fiber Modes

Optical fibers are commonly used for detection systems (Ip et al. 2008). The mode analysis of optical fibers (Marcuse 1974) is similar to the analysis previously discussed on symmetrical waveguides. For cylindrical symmetry, fiber modes (LP_{nm} modes) are defined by the following electromagnetic fields:

$$\mathbf{E}_F = \hat{x} A e^{-j\beta z} \cos(m\psi) \begin{cases} \dfrac{J_m(ur)}{J_m(ud)} & \text{for r} \le \text{d}, \\ \dfrac{K_m(vr)}{K_m(vd)} & \text{for r} > \text{d}, \end{cases}$$

$$\mathbf{H}_F = \frac{\hat{z} \times \mathbf{E}_F}{\eta_0 / n_1}, \tag{2.29}$$

where we assume the electromagnetic field is polarized along the x-direction, n_1 is the refractive index of the fiber core, n_2 is the refractive index of the fiber cladding (Fig. 2.10a), d is the radius of the fiber core, $m = 0,1,2,\ldots$, J are the Bessel functions of the first kind, K are the modified Bessel functions of the second kind, and u and v are the propagation constants defined as

$$u = \sqrt{n_1^2 k_0{}^2 - \beta^2},$$
$$v = \sqrt{\beta^2 - n_2^2 k_0{}^2}. \tag{2.30}$$

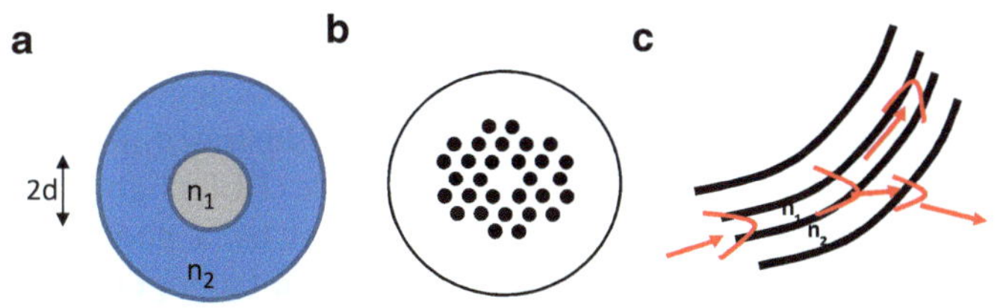

Fig. 2.10 (**a**) Cross section of classic optical fiber. (**b**) Cross section of photonic crystal fiber. (**c**) Emission of light through a fiber curve

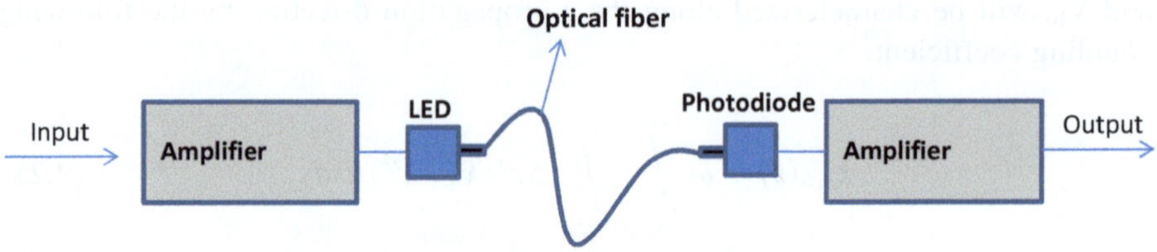

Fig. 2.11 Scheme of a basic communication system using optical fiber

The eigenvalue equation (characteristic equation) is derived by matching the field components at the core/cladding interface and is given by

$$\frac{J_{m-1}(ud)}{uJ_m(ud)} = \frac{K_{m-1}(vd)}{vK_m(vd)}. \tag{2.31}$$

Different kinds of single-mode/multimode optical fiber are available on the optoelectronic market. A particular kind of fiber is the photonic crystal fiber (PCF) (Russel 2006) consisting of a central irregular region defined by multiple air holes (see cross section of Fig. 2.10b). PCFs are divided into two classes. The first one, index-guiding PCF, guides light by total internal reflection between a solid core and a cladding region with multiple air holes. The second class of PCFs uses a 2D periodic structure exhibiting a photonic bandgap (PBG) in order to guide light in a low index core region. Using PCFs, highly birefringent fibers can be easily realized because the index contrast is higher than in conventional fibers. The birefringent behavior can be estimated by the extinction ratios ζ defined as

$$\zeta = 10\log\left(\frac{A_{MIN}}{A_{MAX}}\right), \tag{2.32}$$

where A_{MIN} and A_{MAX} are the minimum and maximum measured optical intensities, respectively, defined by the rotation of a beam splitter. PCFs are also used for biosensing using a proper functionalization or a plasmonic resonance occurring on the core surface. Bending the optical fiber (Taylor 1984) it is possible to provide light outside the core as lost energy. In Fig. 2.10c is shown as a curvature of an optical fiber could emit through as curvatu re. This light emission is used in photonics in order to couple energy with photonic crystal micro-cavities or passive micro/nano components which cannot be touched. A simple example of an optoelectronic telecommunication system is illustrated in Fig. 2.11, where an optical fiber is connected by a light-emitting diode (LED) light source, the light at the output is converted into an electrical signal by a photodiode, and two electronic amplifiers are used to enhance the signal at the input and at the output of the system. Another important application is the plasmonic fiber sensor (Sharma et al. 2007), where a metallic film placed on the core generates an

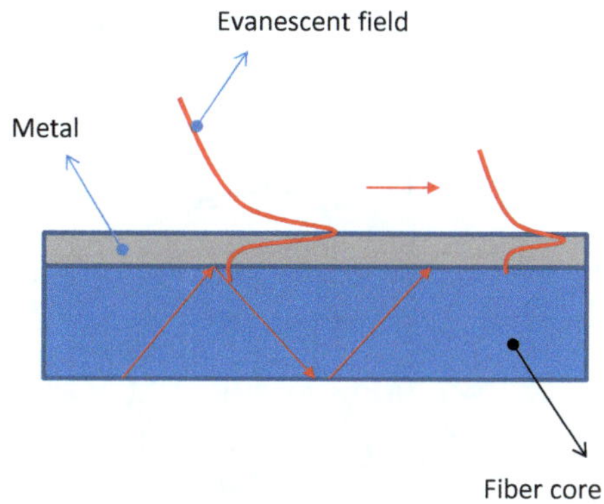

Fig. 2.12 Scheme of plasmonic surface wave propagating along a metallic film placed on a fiber core region

evanescent field (Fig. 2.12) that is able to detect a possible target at the output. The evanescent field will decay as

$$V(z) = V_0 e^{-z/d}, \tag{2.33}$$

where d is the decay length.

2.4 Introduction to the Design and Modeling of Photonic Circuits Integrated Into Optoelectronic Systems

The design and modeling of optical devices are important aspects for the fabrication of efficient optical systems. In this section we will provide some generic aspects concerning the modeling of optical micro devices by highlighting the main aspects and criteria that are useful for a good simulation setting. All the electromagnetic simulators (based on, for example, FDTD, FEM) embed the device into a 3D spatial domain where the calculus is improved. The 3D domain is divided into elementary cells called meshes, which can be rectangular, cubic, or tetrahedral (Fig. 2.13). The calculus is computed in the edge points. For this reason it is necessary that the meshes not be too big in order to ensure that the structure is simulated with a good approximation. As shown in the Fig. 2.13, if the dimension of the structure is small, a fine mesh is required. A big mesh would not be able to "enclose" a thin thickness: in this case, a refinement mesh process would be necessary to improve the convergent solutions.

To decrease computational costs, some intelligent approaches discretize large homogeneous regions (characterized by a unique dielectric constant) using big meshes and inhomogeneous regions (characterized by small parts with a dielectric contrast) using smaller meshes (subgrating process). Among other methods, to decrease

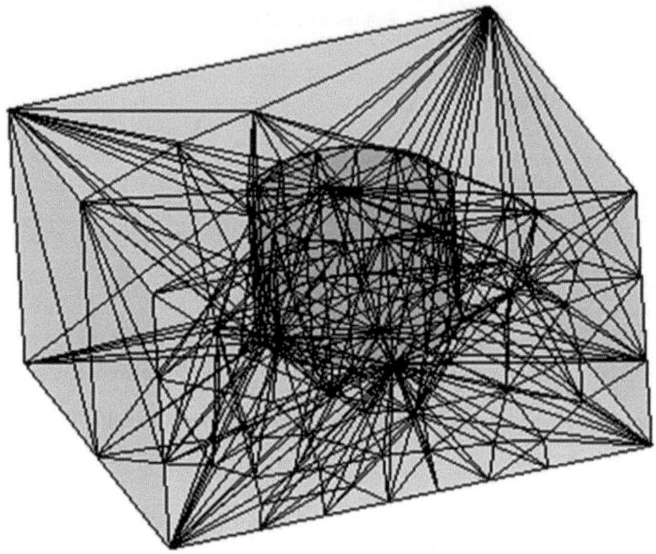

Fig. 2.13 Example of tetrahedral mesh

computational cost, parallel computing can be used to compute small particles/ structures embedded in a large spatial domain (aspect ratio problem). The 3D domain is delimited by ports (facets of the 3D domain) that can distinguish the external environment from the structure (input and outputs) as a black box. The input source excites the device from a port called the input port, and the optical response is provided by the output signal through the output port. A frequency domain code (such as FEM) will provide a frequency response, besides a time domain (such as FDTD), it will provide a time domain signal. If we have a time domain signal to obtain a broad frequency response, a narrow Gaussian pulse should be considered at the input port: a discrete Fourier transform (DFT) will transform the time domain output signal to a frequency signal. In Fig. 2.14 we show the time domain field (Fig. 2.14a) and the DFT (Fig. 2.14b) for a periodic optical waveguide: the irregular trend of the time domain field is due to the reflection of the step discontinuities. In all cases, the mesh dimension must be of an order comparable with that of the working carrier or the central wavelength. The 3D domain box will be characterized by boundary conditions that make it possible to avoid numerical error due to a signal back scattered by the box boundaries: as happens for real devices, the signal coming from an input source will travel along the optical waveguide and flow through the output port. The ports will simulate the presence of air as the external environment. To this end, a perfect matching layer (PML) will simulate the absorbing behavior of the ports and an output signal traveling into the air. PMLs are complex and anisotropic (Berenger 1994). If the device is delimited by metallic layers (as in optoelectronic devices), the metallic boundary condition will be considered. Metallic boundary conditions are ports where the electromagnetic field (solution of Maxwell's equations) is zero

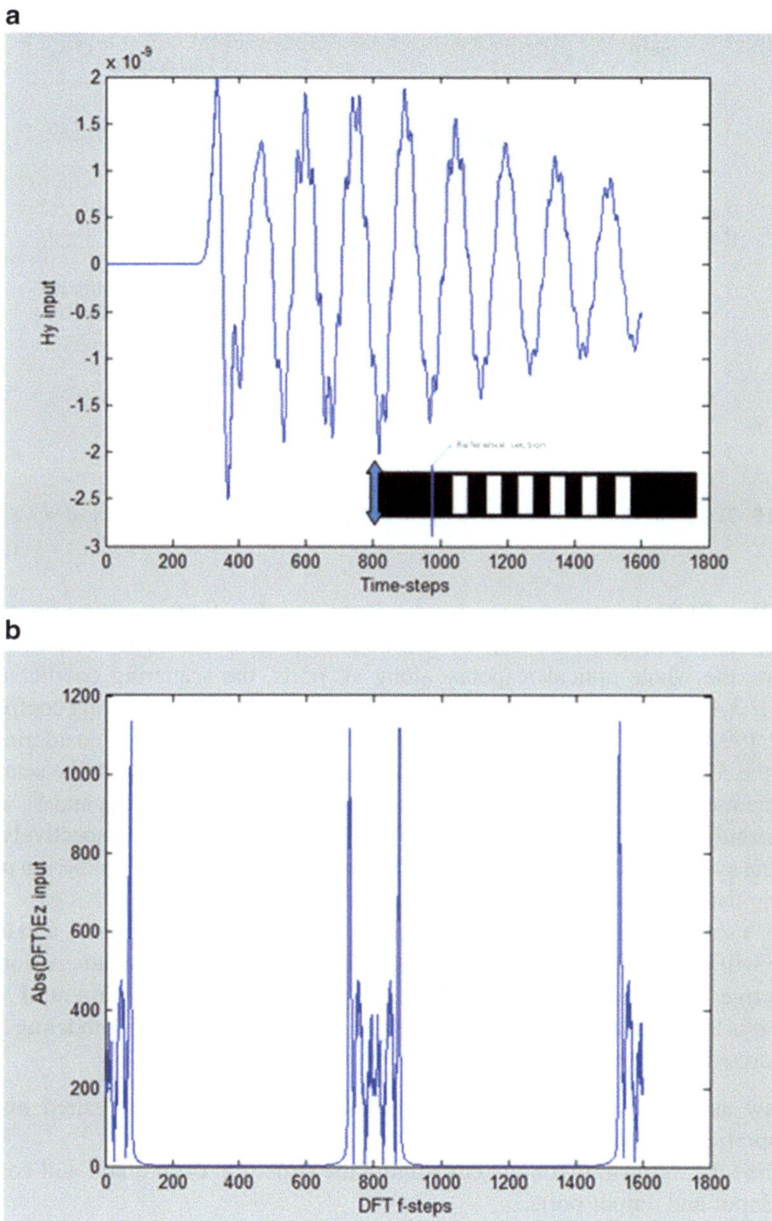

Fig. 2.14 (**a**) Example of time domain magnetic field calculated at input of 2D periodic structure (see inset representing the 2D structure as air holes in a material with $\varepsilon_r = 11$). (**b**) Discrete Fourier transform of electric calculated at the output of the waveguide

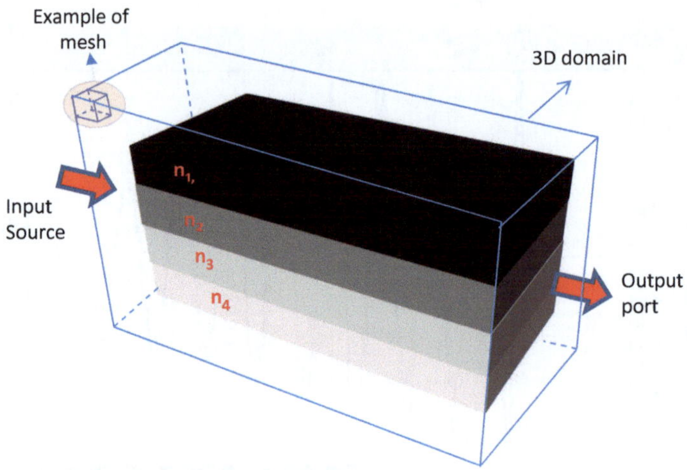

Fig. 2.15 3D spatial domain enclosing a 3D multilayered dielectric passive optical waveguide

($\mathbf{E} = \mathbf{H} = 0$). If the structure is periodic along one direction, it will be possible to decrease the computational costs by defining the periodic boundary conditions. To simulate the whole optical response along all ports, the scattering coefficients S_{ij} ($i = 1,2,3,4,5,6$ and $j = 1,2,3,4,5,6$) should be considered. The scattering coefficients (Pozar 1998) will provide all the reflectivity and the transmittivity considering two ports: the 3D domain will be enclosed in a box with six ports and with the scattering coefficients S_{ii} representing the reflectivity calculated at port i, and S_{ij} and S_{ji} will be the transmittivity observed by exciting the j port and the i port, respectively. The radiating systems, as optical antennas, will also be characterized by a radiation pattern defining the radiation direction (radiation lobes) of the structure. In this case, it is a sphere: a small sphere surrounding the structure will define the near field, and a large sphere will define the far field. Finally, the interactions between metallic regions and conductive materials will be characterized by the density of the current, $\mathbf{J} = \sigma\mathbf{E}$, where σ is the conductivity of the material to be simulated. A correct modeling can be summarized as follows:

1. Draw accurately the structure to simulate implementing the desired material properties (e.g., dielectric permittivity, conductivity).
2. Define the 3D spatial domain embedding the structure: the domain will consider all input and output ports.
3. Define the mesh size according to the working wavelengths to obtain a convergent solution.
4. Check the computational cost according to the PC being used.
5. Restart the simulation by refining the mesh size and compare the results to increase the solution accuracy (Fig. 2.15).

2.4.1 Introduction to Finite Difference Time Domain Method

The finite difference time domain (FTDT) method is based on the finite difference concept. The finite different method discretizes Maxwell's equations in both the time and spatial domains implementing Yee's algorithm (Taflove and Hagness 2000), which discretizes the spatial domain by a rectangular grid. The FDTD method considers rectangular pulses as base functions in both the time and spatial domains. One of the critical issues is the stability of the algorithm. The stability condition for the FDTD method must satisfy the following condition:

$$\Delta t \leq \frac{1}{c\sqrt{\frac{1}{\Delta x^2}+\frac{1}{\Delta y^2}+\frac{1}{\Delta z^2}}}, \tag{2.34}$$

where Δx, Δy, and Δz are the spatial steps of the 3D domain, and Δt is the time step. The FDTD method is suitable for open multicore processors (MPs) and for MPI techniques.

2.4.2 Introduction to Method of Moments

The method of moments (MoM) is suitable for solving electromagnetic problems such as radiation and scattering problems (Harrington 2001; Peterson and Mittra 1997). MoM uses an operator equation. A set of matrix equations is generated by performing symmetric/scalar between the operator and a set of selected testing functions. A typical operator equation can be written as

$$Lg = f, \tag{2.35}$$

where L is a linear operator, f a known function, and g an unknown function that will be solved for. The MoM is based on the following steps:

- Select a basis function expressed as

$$g(t) \approx \sum_{n=1}^{N} p_n g_n(t), \tag{2.36}$$

  where p_n are coefficients to solve. Using (2.35) in (2.36) we obtain the following equation:

$$\sum_{n=1}^{N} p_n L g_n(t) \approx f(t). \tag{2.37}$$

- Select a set of weight functions q_1, q_2,.... q_m to generate a set of linear equations:

$$\sum_{n=1}^{N} Y_{mn} L p_n(t) = k_n \ \ \mathrm{m} = 1, 2, 3, \ldots.\mathrm{N}, \tag{2.38}$$

where

$$Y_{mn} = \int q_n(t) L g_n dt, \tag{2.39}$$

$$k_n = \int q_n(t) f(t) dt, \tag{2.40}$$

and the function $g(t)$ when the coefficients p_{n} will be solved.

2.4.3 Introduction to Finite-Element Method

The finite-element method (FEM) is suitable for solving electromagnetic problems such as closed spaces (e.g., waveguide, cavities). The two following main approaches are used (Rahman et al. 1991; Jin 2002).

- *Scalar formulation.* For example, to calculate the potential ϕ generated by a charge ρ distributed in a domain, we need to solve the Poisson equation

$$\nabla \bullet (\varepsilon \nabla \phi) = -\rho. \tag{2.41}$$

  To find the nodal fields, we must solve the matrix eigenvalue equation

$$T\phi = c, \tag{2.42}$$

  where T is an N × N symmetric matrix, ϕ is an N × 1 unknown vector, and c is a known vector (defined by the charge and by the boundary conditions).
- *Vector formulation.* For example, to calculate the electromagnetic field in a spatial domain, it necessary to solve the following equation:

$$\nabla \times \left(\frac{1}{\mu_r} \nabla \times \mathbf{E} \right) - k_0^2 \varepsilon_r \mathbf{E} = A. \tag{2.43}$$

  Applying the boundary conditions, the equation to solve becomes

$$\mathbf{TE} = \mathbf{c}, \tag{2.44}$$

  where **E** is unknown, **T** is a matrix that implements (2.43), and **c** is a known vector (defined by the charge and by the boundary conditions).

2.4.4 Introduction to Monte Carlo Approach to Opto/Devices

The Monte Carlo (MC) method is associated with a whole class of numerical methods, where the use of a random number generator plays the central role. The MC method, together with the molecular dynamics method, is the main method of computer simulation, especially in the investigation of systems in the thermal equilibrium state. In general cases, all MC methods reduce to the same scheme. That is, we need to calculate the value of a variable c. To do this, we must invent a random variable ψ such that the mathematical expectation of ψ will coincide with the value of c:

$$c = \langle \psi \rangle = P\{\psi\}. \tag{2.45}$$

For a finite N the computational error of the MC method will always be proportional to $N^{-1/2}$. Consequently, the length of the sequence $\{\psi_N\}$ must be very long, and

$$P\{\psi\} = \frac{1}{N}\sum_{i=1}^{N} \psi_{N\to\infty}. \tag{2.46}$$

Actually, the MC method has found wide application only with the advent of fast electronic computers. Metropolis and Ulam (1949) were the first use this method. Evidently, to explore the MC method, first we need a random number generator. Here let us consider the so-called standard random number generator (SRNG), and then we will show how to obtain random numbers (RNs) with any given probability distribution p(ψ) from standard random numbers (SRNs). SRNs are RNs γ that are uniformly distributed within the unit interval (0,1). The sequence of SRNs $\{\gamma_n\}$ must satisfy the following four requirements:

1. Its probability distribution must be equal to one, p(γ).
2. Its mathematical expectation should be $M\{\gamma\} \equiv <\gamma> = 1/2$.
3. The dispersion must be equal to $D\{\gamma\} \equiv <(\gamma - <\gamma>)^2> = 1/12$.
   Recalling that, by definition, the dispersion for a function T(ψ) of a random variable ψ which is defined in an interval (a; b) with a probability distribution p(ψ) is equal to

$$D\{T\} = \left\langle (T - \langle T \rangle)^2 \right\rangle = \int_a^b d\psi p(\psi)[T(\psi) - M\{T\}]^2, \tag{2.47}$$

where

$$M\{T\} = \langle T \rangle = \int_a^b d\psi p(\psi) T(\psi). \tag{2.48}$$

4. There must be no correlation between different elements of the sequence $\{\gamma_n\}$, i.e., the relationship

$$\langle(\gamma_i - 1/2)(\gamma_k - 1/2)\rangle = \langle\gamma_i - 1/2\rangle\langle\gamma_i - 1/2\rangle = 0 \tag{2.49}$$

should be satisfied for any $i \neq k$.

How do we obtain RNs with a computer? The decisive factor here is the fact that all applications of the MC method to physical problems require a very long sequence of RNs. Consequently, such methods of obtaining RNs as, for example, from a table prepared in advance or the setting of the RNs with an external noise generator, turn out to be ineffective (a large table cannot be stored, and with a noise generator it is too difficult to control the quality of the obtained RNs).

The only realistic method turns to be the calculation of RNs with the recurrence first-order formula $\gamma_{n+1} = f(\gamma_n)$. Clearly, in this way we can obtain only pseudo-RNs, not true RNs. The sequence $\{\gamma_n\}$ always has a finite period simply as a consequence of the fact that a set of various numbers available with a computer is finite. Therefore, if some number in the sequence $\{\gamma_n\}$ is encountered a second time, then all subsequent numbers will be repeated too. Thus, the problem is to find such a function $f(\gamma)$ that the period of the obtained sequence $\{\gamma_n\}$ will be as long as possible.

2.5 Wave Propagation and Optical Circuit Concept

Optical modes and plane waves can be modeled by a transmission line (Rozzi and Mongiardo 1997). In this transmission line, where a wave solution of the Helmholtz equation travels in a longitudinal direction (represented in Fig. 2.16 by the z-axis of Fig. 2.1a). The optical source is sketched by a voltage generator, the voltage V signal represents the electric field **E**, and the current signal indicates the magnetic field **H**. Applying resonance conditions, this circuital approach is used to evaluate the modal propagation constants providing dispersion equations. The optical properties of the material where the wave is propagating are implemented in the impedance of the transmission line. By applying proper boundary conditions such as metallic conditions or open waveguide conditions, it is possible to simplify the circuit by an open stub or a short circuit. Closing the circuit by an impedance,

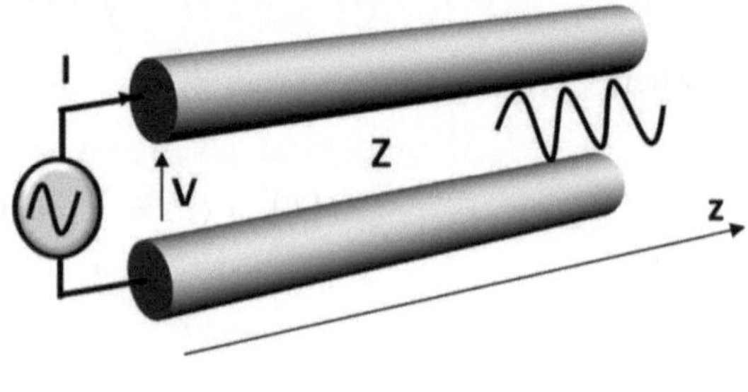

Fig. 2.16 Transmission line model of propagating mode

it is possible to represent the air or substrate region characterized by different propagation constants. Considering a multimode propagation condition in a slab waveguide, each transmission line will represent a propagating TE/TM mode. If a grating couples guided modes, then a coupling admittance displaced in a parallel configuration will model the coupled optical energy (Massaro et al. 2004). Due to the possibility of evaluating a singular electromagnetic field near a discontinuity, the transmission line model is suitable for solving the radiation problem of a 3D dielectric corner (Massaro et al. 2009) and the radiation of metallic plasmonic probe sensors (Massaro and Cingolani 2010).

References

Artigas, D., Rafailov, E.U., Loza-Alvarez, P., Sibbett, W.: Periodically switched nonlinear structures for frequency conversion: theory and experimental demonstration. IEEE J. Quant. Electron. **40**(8), 1122–1130 (2004)

Berenger, J.-P.: A perfectly matched layer for the absorption of electromagnetic waves. J. Comput. Phys. **114**, 185–200 (1994)

Harrington, R.F.: Time Harmonic Electromagnetic Fields. Wiley-IEEE Press, New Jersey (2001)

Ip, E., Tao Lau, A.P., Barros, D.J., Kahn, J.M.: Coherent detection in optical fiber systems. Opt. Expr. **16**(2), 753–791 (2008)

Jin, J.M.: The Finite Element Method in Electromagnetics, 2nd edn. Wiley, New York (2002)

Marcuse, D.: Theory of Dielectric Opt. Waveguides. Academic, New York (1974)

Massaro, A., Cingolani, R.: STRD and near field modeling of metallic wedges for optical detection systems. IEEE J. Light. Technol. **28**(24), 3562–3568 (2010)

Massaro, A., Pierantoni, L., Rozzi, T.: Accurate analysis and modeling of laminated multilayered 3-D optical waveguide. IEEE J. Quant. Electron. **40**(10), 1478–1489 (2004)

Massaro, A., Pierantoni, L., Cingolani, R., Rozzi, T.: A new analytical model of diffrac-tion by 3D-dielectric corners. IEEE Trans. Antennas Propag. **57**(8), 2323–2330 (2009)

Metropolis, N., Ulam, S.: The Monte Carlo method. J. Am. Stat. Assoc. **44**, 335 (1949)

Peterson, A., Mittra, R.: Computational Methods for Electromagnetics. Wiley-IEEE press, New Jersey (1997)

Pozar, D.M.: Microwave Engineering, 2nd edn. Wiley, New York (1998)

Rahman, B., Fernandez, F., Davies, J.: Review of finite element methods for microwave and optical waveguides. Proc. IEEE **79**(10), 1442–1448 (1991)

Rozzi, T., Farina, M.: Advanced Electromagnetic Analysis of Passive and Active Planar Structures. IEE Electromagnetic Wave Series, vol. 46. Institution of Electrical Engineers, London (1999)

Rozzi, T., Mongiardo, M.: Open Electromagnetic Waveguides. IEE Electromagnetic Waves Series, vol. 43. Institution of Electrical Engineers, London (1997)

Russel, P.S.J.: Photonic-crystal fibers. IEEE J. Light. Technol. **24**(12), 4729–4749 (2006)

Sharma, A.K., Jha, R., Gupta, B.D.: Fiber-optic sensor based on surface plasmon resonance: a comprehensive review. IEEE Sens. J. **7**(8), 1118–1129 (2007)

Taflove, A., Hagness, S.C.: Computational Electromagnetic: The Finite-Difference Time-Domain Method, 2nd edn. Artech House, London (2000)

Taylor, H.F.: Bending effects in optical fibers. IEEE J. Light. Technol. **2**(5), 617–628 (1984)

Venugopal, R.S., Moutzouris, K., Ebrahimzadeh, M.: Nonlinear frequency conversion in semiconductor optical waveguides using birefringent, modal and quasi-phase-matching technique. J. Opt. A: Pure Appl. Opt. **6**, 73241–73249 (2004)

Chapter 3
Optical and Micro-/Nanosensing

Abstract This chapter presents an overview of optical and wireless devices oriented toward sensing approaches. Optical fiber implementations, distributed Bragg reflectors, and plasmonic waveguides, including micro-/nanosensors, are discussed. A particular view is addressed on modeling and applications.

3.1 Optical Sensors and Devices

The monitoring process of activities is actually an important issue in many fields. The development of sensors is a key issue in many activities, especially in such areas as military, aerospace, renewable energy, biotechnology, nanotechnology, robotics, and others. Efficient sensor implementation and characterization are needed to optimize sensor processes for environmental, industrial, and bio monitoring. In particular, optical fiber layouts and optical waveguides are commonly used because they are easy to implement. In applications such as oil spills, robotic tactile sensors, pollution detection, optical permittivity sensing, plasmonic waveguiding, and optical antennas represent alternative sensing approaches compared to traditional sensing approaches using electrical connections and circuits.

3.1.1 Examples of Optical Fiber Sensors for Environment Monitoring

As a first example of an optical detection system, we discuss in this section an oil spill optical fiber sensor. Oil spills in water can seriously affect the environment and, consequently, human health. An example implementing optical fibers is illustrated in Fig. 3.1, where an input transmitter optical fiber embedded in water excites the water/air interface. The light is then reflected by this interface and coupled to a

A. Lay-Ekuakille, *Optical Waveguiding and Applied Photonics*, Lecture Notes in Nanoscale Science and Technology 10, DOI 10.1007/978-1-4614-5959-0_3,

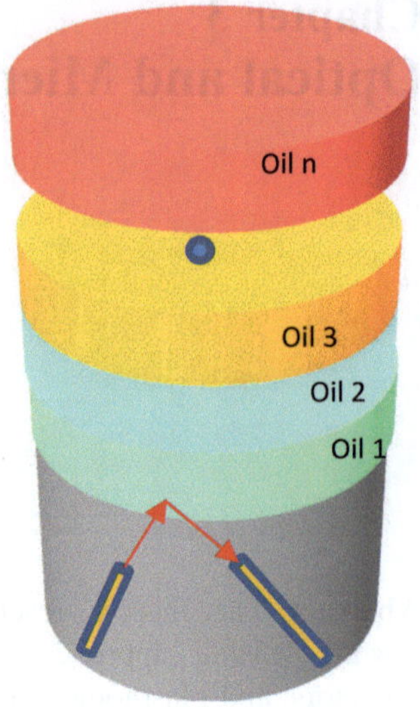

Fig. 3.1 3D design of optical antenna sensor. The two transmitter and receiver optical fibers are embedded in a water volume. Oil_1, oil_2, oil_3,... oil_n could represent the oil droplets placed in water

receiver optical fiber embedded in the same water volume. The coupled light will increase with oil concentration (the oil concentrations satisfy the following condition: $oil_1 < oil_2 < oil_3 < \ldots oil_n$) because the air/interface behaves as an optical mirror. For a large extension of oil on the surface, the optical system could be supported by a proper image postprocessing able to detect the area to measure.

Another example of optical fiber implementation is the system illustrated in Fig. 3.2, which can detect micropollutants (in this case hydrocarbons) in a water solution. Also, in this case an input optical fiber could excite the sample, and an optical receiver could collect the light at the output of the sample. The pollution of a liquid is characterized by microparticles floating in the water solution. Due to the small dimensions, the microparticles interact with light: the light coming from an external broad source can be absorbed and scattered inside the impure liquid.

This basic principle explains how it is possible to detect, in real time, the pollution of a water sample with floating hydrocarbons. As observed in Fig. 3.2a, when a light source excites a fixed sample (without motion), small quantities of microparticles will be floating inside the water, and consequently, a large quantity of light will be transmitted at the output of the sample (corresponding to a small quantity of absorbed and scattered light of the liquid). When a shaker is applied to the sample, large quantities of initially deposited hydrocarbon microparticles will be uniformly dispersed throughout the volume of the sample, and a large quantity of the light will be absorbed by the decreased intensity of the transmitted light. If no vibrations are displayed, and increased a low variation of the transmitted optical intensity is observed. This variation becomes strong if we apply a shaker. The effect of the shaker is to decrease the transmitted light collected by an output receiver optical fiber probe, as observed by the theoretical trend of Fig. 3.2b. In both examples shown, an optical analyzer could be connected to the output fiber to

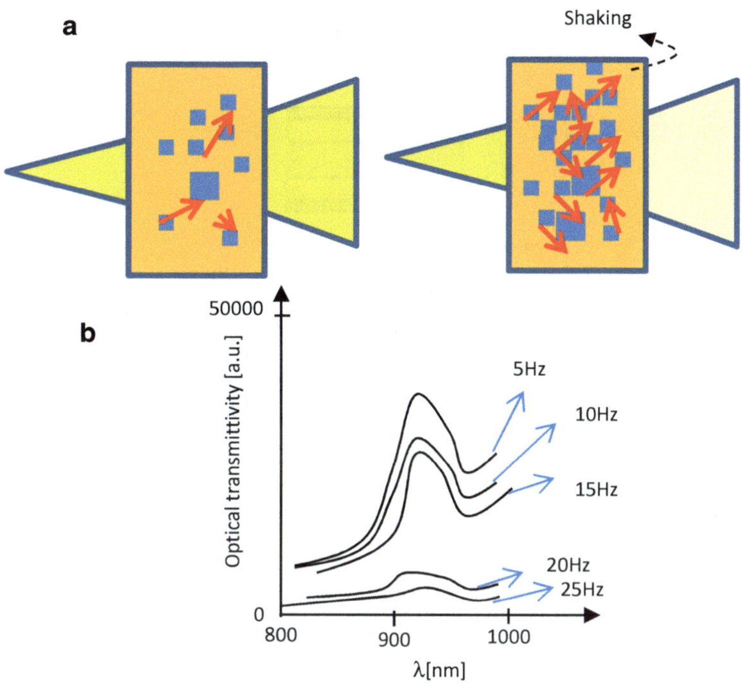

Fig. 3.2 Schematic illustration of light scattered by micropollutants in water. The light source region is defined by the light cone generated by an input fiber source. The output fiber should collect the light at the output of the sample (an optical lens could be used to better focus the output light on the core of the output optical fiber). (**a**) Light scattering (*left*) of a liquid sample with hydrocarbons (are not improved vibrations) and light scattering (*right*) of the vibrated sample. (**b**) Trend of spectra of hydrocarbon sample at different vibration frequencies

measure the optical spectra. An optical lens system could be used in the second case to better focus the output light into a spectrum analyzer. The spectrum analyzer could be substituted into a photodetector connected to a digital oscilloscope. The receiver and the transmitter optical fibers could also be used for gas sensing.

3.2 Distributed Bragg Reflectors

A particular kind of optical filter is the distributed Bragg reflector (DBR). The layout of a DBR follows the scheme of Fig. 3.3, where a couple of dielectric layers are placed above and below a central microcavity. The entire multilayer structure is grown on a buffer layer deposited on a substrate (e.g., glass). The refractive index n and the thicknesses d must satisfy the following condition: $n_1 d_1 = n_2 d_2 = \lambda_0/4$, $n_c d_c = \lambda_0/2$, where λ_0 is the central wavelength. An example of DBR takes into

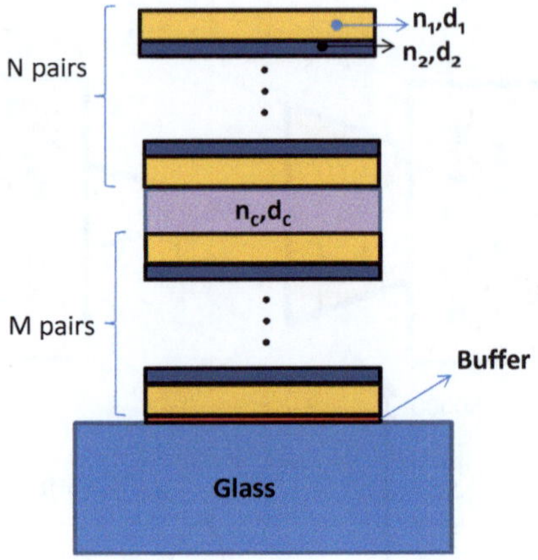

Fig. 3.3 (**a**) Typical configuration of a GaN/AlGaN distributed Bragg reflector (DBR). (**b**) DBR reflectance R: comparison of experimental and numerical (FEM) results

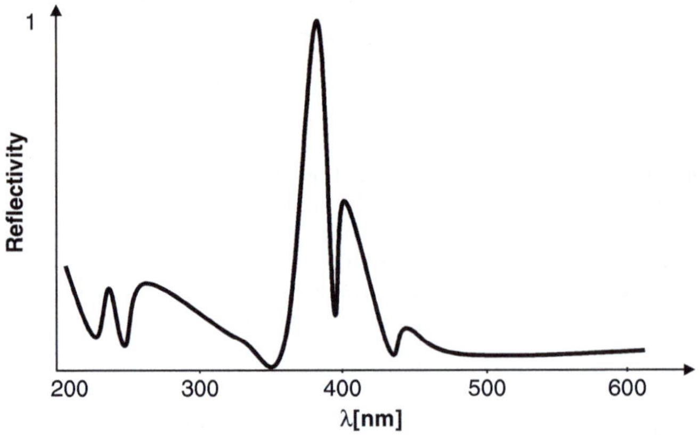

Fig. 3.4 FEM theoretical trend of reflectivity response for a GaN/AlGaN distributed Bragg reflector

account nitride materials for nonlinear applications. In particular, such materials show a second-order nonlinear optical response $\chi^{(2)}$ comparable to conventional nonlinear crystals such as KTP or $LiNbO_3$. For example, we simulate and measure a GaN microcavity embedded between two DBRs consisting of AlGaN/GaN multi-layers on a sapphire substrate that is designed to enhance the cavity resonance around $\lambda_0 = 400$ nm (see the theoretical FEM reflectivity trend in Fig. 3.4, where five GaN/AlGaN couples above and below the central GaN microcavity are considered, respectively. The DBR can be designed to resonate (resonance of the central microcavity) around $\lambda_0 = 400$ nm. In this case, a second harmonic (SH) signal generated at $\lambda = 800$ nm (an important aspect useful for multiplexing systems) can be extracted from the cavity to be guided into ridge waveguides.

3.3 Laser: Active Materials and Modeling

Vertical microcavity lasers based on simple Fabry Perot cavities having high-reflectivity dielectric quarter-wave stacks for mirrors have attracted much attention because of their desirable lasing characteristics, in particular, for their low-power consumption, high-output power, and single-mode operation. The low-lasing threshold of VCELs can be further improved using a periodic gain configuration within the cavity. By placing thin gain segments along the **E**-field standing-wave maxima, the longitudinal confinement factor is maximized, thereby reducing the material gain threshold. Figure 3.5a, b illustrates the frequency response and the propagated pulse of a multilayer stack embedding an active central cavity, respectively. The cavity material is modeled by the following complex conductivity:

$$\sigma(\omega) = \sigma_1(\omega) + j\sigma_2(\omega). \tag{3.1}$$

Assuming that the propagating wave is in the form (x-propagating direction)

$$V_z(x,t) = V_0 e^{(-\alpha x)} e^{[-j(\beta x - \omega t)]}, \tag{3.2}$$

the effective relative permittivity will be

$$\varepsilon_{eff}(\omega) = \varepsilon_r - \frac{\sigma_2(\omega)}{\omega\varepsilon_0} \tag{3.3}$$

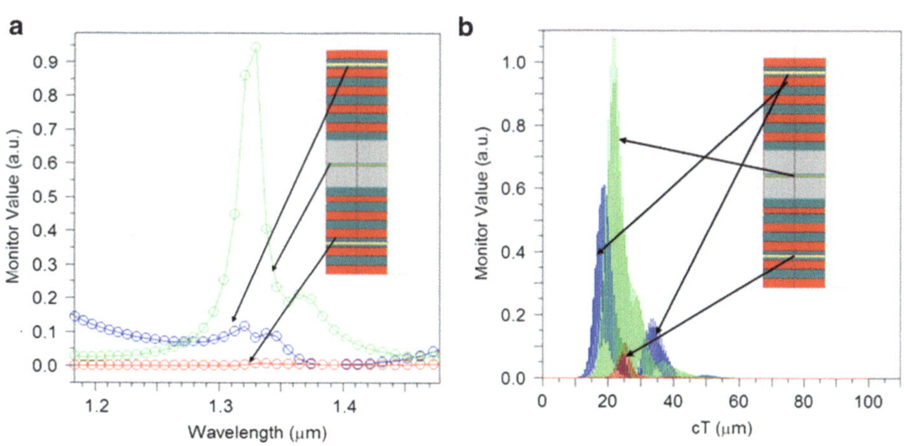

Fig. 3.5 (**a**) Example of active central cavity in GaN/AlGaN multilayer stack configuration. The cavity is characterized by a gain around $\lambda = 1.33$ μm. (**b**) Time evolution of pulse propagated in different regions (numerical monitors)

and

$$\alpha(\omega) \cong \frac{\sigma_1(\omega)}{2c\varepsilon_0 n_{eff}}, \tag{3.4}$$

$$\beta(\omega) \cong n_{eff}\frac{\omega}{c}, \tag{3.5}$$

where $n_{eff}^{\ 2} = \varepsilon_{r,eff}$. From (3.19) we conclude that if $\sigma_1(\omega)$ is negative, then α is negative, providing a gain.

3.4 Integrated Optical Gratings

A single-mode and polarization-maintaining, planar-type, three-dimensional (3D) dielectric waveguide is a goal to achieve in the integration of optical devices. Integrated optical gratings could consist of a very thin film lamination a-S_i:H, S_iO_2 of layers having different refractive indices, which alternate periodically through high and low values (see different cases shown in Fig. 3.6a). This kind of waveguide could support a single longitudinal section electric (LSE) mode and a

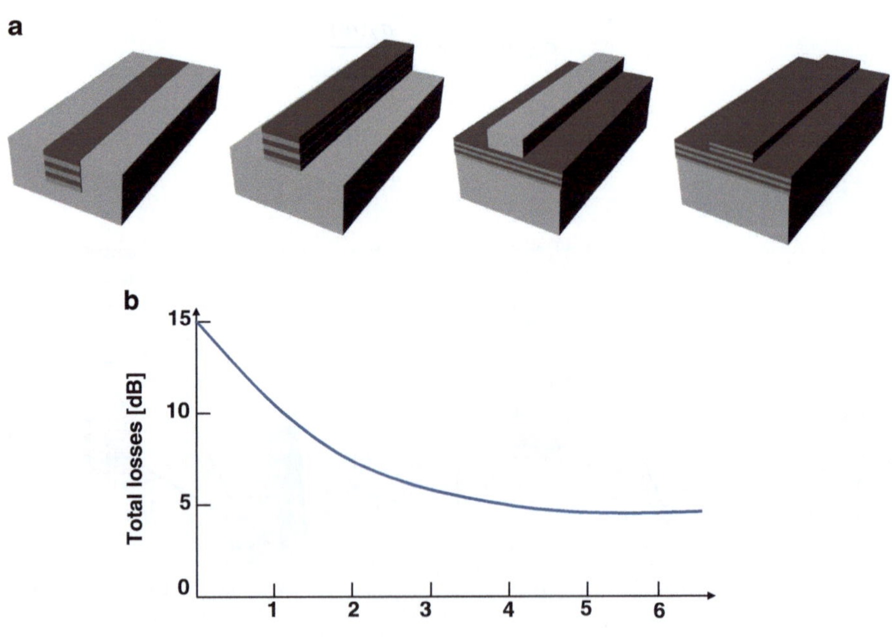

Fig. 3.6 (**a**) Possible layouts of integrated dielectric multilayered waveguides. (**b**) Theoretical trend of total losses for a multilayered waveguide versus rib width

single longitudinal section magnetic (LSM) mode. These modes are coupled by the presence of grating along the longitudinal *z*-axis. Using the effective dielectric constant (EDC) method, it is possible to determine the geometrical parameters of the rib waveguide for a single LSE and a single LSM propagation mode at a wavelength of 1.55 μm (telecommunication working wavelength). This single-mode condition is useful for improving the maximum guided energy (and consequently to reduce losses). Figure 3.6b shows a theoretical loss trend for a certain multilayered waveguide (ARROW waveguide) versus rib width. The analyzed devices can also be used for integrated DBRs and for electro-optic switches.

3.5 Nonlinear Optics: Second Harmonic Generation Processes and Design of χ^2 Waveguides

Second harmonic generation processes in $\chi^{(2)}$ nonlinear waveguides (Venugopal et al. 2004) are present in the GaAs/AlGaAs nonlinear waveguide with dielectric discontinuities shown in Fig. 3.5a. The quasi-phase-matching (QPM) condition can be applied to evaluate a good SH conversion efficiency. The QPM technique is a practical method for substantially increasing the SH power by effectively reducing the phase mismatch between the fundamental and SH fields: a properly designed grating compensates the difference of the effective refractive indexes associated with the fundamental and SH propagating modes. The QPM application in high-efficiency SH generation processes is analyzed in this section. For example, we model the $\chi^{(2)}$ nonlinear process in the asymmetrical GaAs slab waveguide with a nonlinear core and dielectric discontinuities: in the nonlinear planar waveguides a fundamental mode ($\lambda_{FU} = 1.955$ μm) is coupled to a copropagating SH mode ($\lambda_{SH} = 0.9775$ μm) through an appropriate nonlinear susceptibility coefficient. Figure 3.7b shows the theoretical SH intensity at the output of the waveguide of Fig. 3.7a.

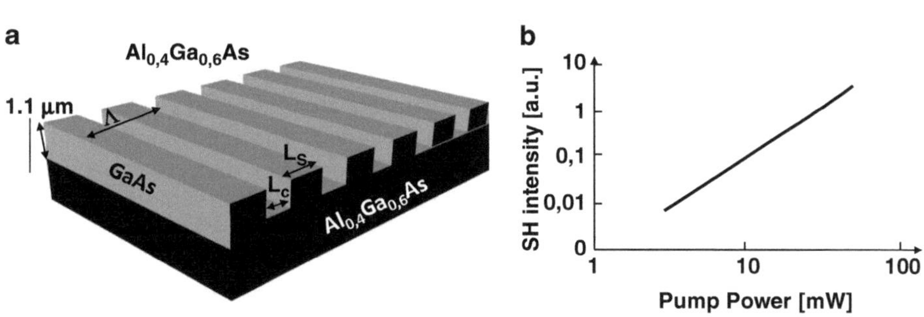

Fig. 3.7 (**a**) GaAs/AlGaAs waveguide obtained by periodically switched nonlinearity (*PSN*). (**b**) Theoretical SH intensity trend estimated at waveguide output

3.6 Plasmonic Waveguides

The concentration of optical energy below the diffraction limit has many diverse applications. They include near-field microscopy, single-molecule biochemical sensing, miniaturized integrated optics, high-density data storage, nonlinear optics, surface-enhanced Raman scattering or fluorescence, and implanted wireless systems. Metallic wedges are suitable for these applications because they can focus the energy near the discontinuity, and they behave as probes for optical detection systems. The optical source could be a surface plasmonic wave that could be enhanced by a nano/micro grating placed directly on the probe. Plasmonic waves (Simon et al. 1975) are generated by properly exciting a planar metal plasmon waveguide.

In this section, we define the theoretical model of the planar metal plasmon waveguide depicted in Fig. 4.14. Assuming as the source an electric field $\mathbf{V_0}$ polarized parallel to the plane of incidence (in accordance with the p-polarized light, which can excite electronic surface plasmons), the resulting surface electromagnetic wave will have the following general form:

$$\begin{aligned} V_1 &= V_{\mathbf{10}} e^{i(\mathbf{k}_{x1}\mathbf{x}+\mathbf{k}_{z1}\mathbf{z}-\omega t)} && x < 0, \\ V_2 &= V_{\mathbf{20}} e^{i(\mathbf{k}_{x2}\mathbf{x}+\mathbf{k}_{z2}\mathbf{z}-\omega t)} && x > 0, \end{aligned} \tag{3.6}$$

where $\mathbf{V}$ stands for $\mathbf{E}$, $\boldsymbol{k_{x1}}$ and $\boldsymbol{k_{x2}}$ are the wavevectors in the x-direction, $\boldsymbol{k_{z1}}$ and $\boldsymbol{k_{z2}}$ are those in the z-direction, and ω is the angular frequency. Both electric and magnetic fields must fulfill the Maxwell equations:

$$\begin{aligned} &\nabla \cdot \mathbf{I} = 0, \\ &\nabla \cdot \mathbf{V} = 0, \\ &\nabla \times \mathbf{V} + \frac{1}{c}\frac{\partial \mathbf{I}}{\partial t} = 0, \\ &\nabla \times \mathbf{I} + \frac{\varepsilon}{c}\frac{\partial \mathbf{V}}{\partial t} = 0, \end{aligned} \tag{3.7}$$

where $\mathbf{I}$ stands for $\mathbf{H}$. The incident light coming from the dielectric material reflects at the metallic interface, inducing light localization, called an evanescent field (EW). The EW excites the SPW which propagates along the surface of the metal film, and thus part of the incident light is absorbed. The light intensity of the reflected ray depends on the incidence angle. The complex dielectric function of a metallic film can be expressed in the following form:

$$\varepsilon_m(\omega) = \varepsilon'_m(\omega) + i\varepsilon''_m(\omega). \tag{3.8}$$

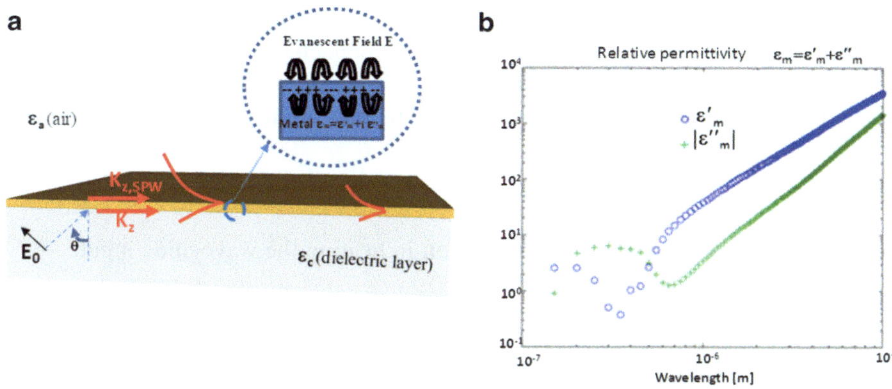

Fig. 3.8 (**a**) Plasmon waveguide and evanescent field **E** = **V** (surface plasmon waves) propagating along *z*-direction. (**b**) Gold permittivity versus wavelength calculated by Brendel–Bormann model

According to the Brendel–Bormann (BB) model, the expression becomes

$$\varepsilon_m(\omega) = 1 - \frac{\upsilon_p^2}{\omega(\omega - i\Gamma_0)} + \sum_{j=1}^{k} S_j(\omega), \qquad (3.9)$$

where k is the number of BB oscillators used to interpret the intraband part of the spectrum, and $\nu_P = (f_0\omega_P)^{1/2}$ is the plasma frequency associated with the intraband transition with oscillator strength f_0 and damping constant Γ_0 and with an infinite number of oscillators (part of which replaces the Lorentz oscillators) given by

$$S_j(\omega) = \frac{1}{\sqrt{2\pi}\sigma_j} \int_{-\infty}^{\infty} \exp\left[\frac{-(t-\omega_j)^2}{2\sigma_j^2}\right] \times \frac{f_j\omega_p^2}{(t^2-\omega^2)+i\omega\Gamma_j} dt. \qquad (3.10)$$

Figure 3.8b reports the calculation of a gold complex permittivity by the BB approach. The propagation constants of the SPW define the electromagnetic field profile and change with the working wavelength following the dielectric permittivity trend.

A simple experimental approach that measures a plasmonic EW can be performed by considering a commercial glass waveguide with indium tin oxide (ITO) as the top layer. In particular, it is possible to quantify the plasmonic EW intensity by attaching an optical fiber detector directly on the ITO surface. An input light probe beam (for example, coming from a tungsten broadband lamp) can be injected into the plasmonic waveguide through a tapered fiber that is not attached to the waveguide to improve the plane-wave source. The light exciting the waveguide and radiated by the metallic ITO layer can be collected by a multimode fiber positioned directly above the ITO layer and connected to a spectrum analyzer. The power transferred by the waveguide will decrease with distance z. For a particular z-position of the multimode fiber, the plasmonic power is estimated as

$$P(z) = \left|V_1(\omega, z)/V_1(\omega, z_0)\right|^2, \tag{3.11}$$

where z_0 represents the reference z-position indicating the waveguide input.

3.7 Monte Carlo Simulation of Nanoparticle Array

3.7.1 Monte Carlo Molecular Modeling

Monte Carlo molecular modeling is the application of Monte Carlo methods to molecular problems. These problems can also be modeled using the molecular dynamics method. The difference is that this approach relies on statistical mechanics rather than molecular dynamics. Instead of trying to reproduce the dynamics of a system, it generates states according to the appropriate Boltzmann probabilities. Thus, it is the application of the Metropolis Monte Carlo simulation to molecular systems. It is therefore also a particular subset of the more general Monte Carlo method in statistical physics. It employs a Markov chain procedure to determine a new state for a system from a previous one.

According to its stochastic nature, this new state is accepted at random. Each trial usually counts as a move. The avoidance of dynamics restricts the method to studies of static quantities only, but the freedom to choose moves makes the method very flexible. These moves must only satisfy a basic condition of balance in order for the equilibrium to be properly described, but a detailed balance, a stronger condition, is usually imposed when designing new algorithms. An additional advantage is that some systems, such as the Ising model, lack a dynamical description and are only defined by an energy prescription; for these the Monte Carlo approach is the only feasible one. The great success of this method in statistical mechanics has led to various generalizations such as the method of simulated annealing for optimization, in which a fictitious temperature is introduced and then gradually lowered. The general motivation to use the Monte Carlo method in statistical physics is to evaluate a multivariable integral. The typical problem begins with a system whose Hamiltonian is known; it is at a given temperature and follows Boltzmann statistics. To obtain the mean value of some macroscopic variable, called A, the general approach is to compute, over the entire phase space, the mean value of A using the Boltzmann distribution:

$$\langle C \rangle = \int_{phase\ space} C_r \frac{e^{-\beta E_r}}{Z} dr, \tag{3.12}$$

where E_r is the energy of the system for a given state defined by **r**, a vector with all the degrees of freedom [for instance, for a mechanical system, $\mathbf{r} = (\mathbf{q},\mathbf{p})$], $\beta = 1/k_bT$, and

$$Z = \int_{phase\ space} P(\mathbf{r})d\mathbf{r} \tag{3.13}$$

is the partition function. One possible approach to solving this multivariable integral is to exactly enumerate all possible configurations of the system and calculate the averages at will. This is actually done in exactly solvable systems and in simulations of simple systems with few particles. In realistic systems, on the other hand, even an exact enumeration can be difficult to implement. For those systems, the Monte Carlo integration (not to be confused with the Monte Carlo method, which is used to simulate molecular chains) is generally employed. The main motivation for its use is the fact that, with the Monte Carlo integration, the error proceeds as $1/\sqrt{N}$, independently of the dimension of the integral. Another important concept related to the Monte Carlo integration is importance sampling, a technique that improves the computational time of the simulation.

3.7.2 Importance Sampling

The estimative, under Monte Carlo integration, of an integral is defined as

$$\langle C \rangle \cong \frac{1}{N}\sum_{i=1}^{N} C_{ri} \frac{e^{-\beta E_{ri}}}{Z} dr, \tag{3.14}$$

where $\boldsymbol{r}_i$ are uniformly obtained from the entire phase space and N is the number of sampling points (or function evaluations). From the entire phase space, some zones are generally more important for the mean of variable A than others. In particular, those that have a $e^{-\beta Eri}$ value that is sufficiently high when compared to the rest of the energy spectra are the most relevant for the integral. Using this fact, the natural question to ask is: is it possible to choose, with greater frequency, the states that are known to be more relevant to the integral? The answer is yes, using the importance sampling technique. Let us assume that p(**r**) is a distribution that chooses the states that are known to be more relevant to an integral. The mean value of A can be rewritten as

$$\langle C\rangle = \int_{phase\ space} p^{-1}(\mathbf{r})\frac{C_r}{p^{-1}(\mathbf{r})}\frac{e^{-\beta E_r}}{Z}dr = \int_{phase\ space} p^{-1}(\mathbf{r})C_r^{r}\frac{e^{-\beta E_r}}{Z}dr, \quad (3.15)$$

where C_r^* are the sampled values taking into account the importance probability p(**r**). This integral can be estimated by

$$\langle C\rangle \cong \frac{1}{N}\sum_{i=1}^{N} p^{-1}(\mathbf{r})C^*{}_{ri}\frac{e^{-\beta E_{ri}}}{Z}dr, \quad (3.16)$$

where $\mathbf{r}_i$ are now randomly generated using the p(**r**) distribution. Since most of the time it is not easy to find a way to generate states with a given distribution, the Metropolis algorithm must be used.

Because it is known that the most likely states are those that maximize the Boltzmann distribution, a good distribution, p(**r**), to choose for importance sampling is the Boltzmann distribution or canonical distribution. Let

$$p(\mathbf{r}) = \frac{e^{-\beta E_{ri}}}{Z} \quad (3.17)$$

be the distribution to use. Substituting into the previous sum,

$$\langle C\rangle \cong \frac{1}{N}\sum_{i=1}^{N} C^*{}_{ri} \quad (3.18)$$

Thus, the procedure for obtaining a mean value of a given variable using the Metropolis algorithm (), with a canonical distribution, is to use the Metropolis algorithm to generate states given by the distribution p(**r**) and perform means over C_{ri}^*.

One important issue must be considered when using the Metropolis algorithm with a canonical distribution: when performing a given measure, i.e., realization of $\mathbf{r}_i$, one must ensure that that realization is not correlated with the previous state of the system (otherwise the states are not being randomly generated). On systems with relevant energy gaps, this is the major drawback of using a canonical distribution because the time needed for the system to decorrelate from the previous state can tend to infinity.

3.7.3 Metropolis–Hastings Algorithm

In statistics and in statistical physics, the Metropolis–Hastings algorithm (Chib and Greenberg 1995), also known as the Metropolis algorithm, is a Markov chain Monte Carlo (MCMC) method for obtaining a sequence of random samples from a probability distribution for which direct sampling is difficult. This sequence can be used to approximate the distribution (i.e., to generate a histogram) or to compute an integral (such as an expected value). Metropolis–Hastings and other MCMC algorithms are generally used for sampling from multidimensional distributions, especially when the number of dimensions is high. For single-dimensional distributions, other methods are usually available (e.g., adaptive rejection sampling) that can directly return independent samples from the distribution and are free from the problem of autocorrelated samples, which is inherent in MCMC methods.

The objective of the Metropolis–Hastings algorithm is to asymptotically generate states x according to a desired distribution P(x). To this end, the Metropolis–Hastings algorithm uses a stochastic process. The idea is to construct a stochastic process that asymptotically converges to P(x), and thus, after a transient period, it generates states according to it. The process is defined as Markovian, on which each new state x' is generated from the current state x, and this transition does not depend on the history of the process. The transition probability can be written as a Markov matrix P($x \rightarrow x'$). The question is how this matrix can be chosen such that, asymptotically, the states are generated according to P(x). It can be shown that two conditions are necessary and sufficient: the ergodicity of the process, which ensures that at most one asymptotic distribution exists, and the condition of balance of the matrix, which is generally satisfied if the matrix satisfies a detailed balance. Detailed balances ensure that there exists at least one asymptotic distribution and that this asymptotic distribution is P(x). Formally, the detailed balance states that the probability of being in state x and transiting to x' must be the same as the probability of being in state x' and transiting to x. This can be written as

$$P(x)P(x \rightarrow x') = g(x')C(x' \rightarrow x), \tag{3.19}$$

which can be rearranged as

$$\frac{P(x \rightarrow x')}{P(x' \rightarrow x)} = \frac{P(x')}{P(x)}. \tag{3.20}$$

One important issue to address is how the dynamics of the process can be set using this transition matrix. One common approach is to separate each step of the process into two substeps: proposal and acceptance refusal. The idea is to first propose a state x' according to a given distribution g($x \rightarrow x'$) and accept it according to C($x \rightarrow x'$). This means that for the system to transit from x to x', the state x' must be both proposed and accepted. These two subprocesses are equivalent to a transition, and so

$$P(x \rightarrow x') = g(x \rightarrow x')C(x \rightarrow x'). \tag{3.21}$$

This relation can be inserted into the previous equation to obtain

$$\frac{C(x \rightarrow x')}{C(x' \rightarrow x)} = \frac{P(x')}{P(x)} \frac{g(x' \rightarrow x)}{g(x \rightarrow x')}. \tag{3.22}$$

This equation is still redundant in the sense that there are several As that lead to this same relationship. A typical choice for the acceptance is

$$C(x \rightarrow x') = \min\left(1, \frac{P(x')}{P(x)} \frac{g(x' \rightarrow x)}{g(x \rightarrow x')}\right), \tag{3.23}$$

which is known as the Metropolis choice. This equation completes the formal derivation of the process. From a practical point of view, the Metropolis–Hastings algorithm consists of the following steps:

1. Choose an initial state x at random.
2. Randomly choose a state x' according to $g(x \rightarrow x')$.
3. Accept the state according to $C(x \rightarrow x')$. If it is not accepted, then $x' = x$, and so there is no need to update anything. Else, the system transits to x'.
4. Go to 2 until T states are generated.
5. Save state x, go to 2.

In principle, this procedure ensures that stored states are not correlated, and T must be chosen based on various factors, such as, for example, the distribution g and the acceptance it leads.

It is not clear which distribution $g(x \rightarrow x')$ one should use; it is a free parameter of the method. However, if one knows nothing, one can always use a uniform distribution. In this case, T can be set to 1 because the proposed state x' is always decorrelated from the present state x.

3.7.4 The Quasi-Monte Carlo Variant

Quasi-Monte Carlo simulation is the traditional Monte Carlo simulation but using quasi-random sequences instead of (pseudo) random numbers. These sequences are used to generate representative samples from the probability distributions that we are simulating in our practical problem. The quasi-random sequences, also called low-discrepancy sequences, in several cases permit one to improve the performance of Monte Carlo simulations, offering shorter computational times or higher accuracy.

In reality, low-discrepancy sequences are totally deterministic, so the popular name quasi-random can be misleading.

Lemieux mentioned the drawbacks of quasi-Monte Carlo:

- In order for

$$O\left(\frac{(\log N)^S}{N}\right) \tag{3.24}$$

to be smaller than

$$O\left(\frac{1}{\sqrt{N}}\right), \tag{3.25}$$

s needs to be small and n needs to be large.

- For many practical functions, $V(f) = \infty$.
- We only know an upper bound on the error (i.e., $\varepsilon \leq V(f)\, D_N$), and it is difficult to compute D_N^* and V(f).

To overcome these difficulties, we can use a randomized quasi-Monte Carlo method. The resulting method is called the randomized quasi-Monte Carlo method and can also be viewed as a variance reduction technique for the standard Monte Carlo method. Among several methods, the simplest transformation procedure is through random shifting. Let $\{x_1, \ldots, x_N\}$ be a point set from a low-discrepancy sequence. We sample s-dimensional random vector U and mix it with $\{x_1, \ldots, x_N\}$. Specifically, for each x_j, create

$$y_j = x_j + U(\text{mod}1) \tag{3.26}$$

and use the sequence (y_j) instead of (x_j). If we have R replications for Monte Carlo, then we sample the s-dimensional random vector U for each replication. The drawback of randomization is the sacrifice of computation speed. Since we now use a pseudorandom number generator, the method is slower. Randomization is useful since the variance and the computation speed are slightly better than that of the standard Monte Carlo method.

The concept of low discrepancy is associated with the property that successive numbers are added to a position as far as possible from the others numbers, that is, avoiding clustering (groups of numbers close to each other). The sequence is constructed based on the idea that each point is repelled from the others. Thus, if the idea is for the points to maximally avoid each other, the job for the numbers generated sequentially is to fill in the larger "gaps" between the previous numbers of the sequence.

3.7.5 *Variant Van der Corput Sequence*

A van der Corput sequence is a low-discrepancy sequence over the unit interval. It is constructed by reversing the base n representation of the sequence of natural numbers (1, 2, 3, ...). For example, the decimal van der Corput sequence begins as follows:

$$0.1, 0.2, 0.3, 0.4, 0.5, 0.6, 0.7, 0.8, 0.9, 0.01, 0.11, 0.21, 0.31, 0.41, 0.51,\\ 0.61, 0.71, 0.81, 0.91, 0.02, 0.12, 0.22, 0.32, \ldots,$$

whereas the binary van der Corput sequence can be written as

$$0.1_2, 0.01_2, 0.11_2, 0.001_2, 0.101_2, 0.011_2, 0.111_2, 0.0001_2, 0.1001_2, 0.0101_2,\\ 0.1101_2, 0.0011_2, 0.1011_2, 0.0111_2, 0.1111_2, \ldots$$

or, equivalently, as

$$1/2, 1/4, 34, 1/8, 5/8, 3/8, 7/8, 1/16, 9/16, 5/16, 13/16, 3/16, 11/16, 7/16, 15/16, \ldots$$

The elements of the van der Corput sequence (in any base) form a dense set in the unit interval: for any real number in [0, 1] there exists a subsequence of the van der Corput sequence that converges toward that number. They are also equidistributed over the unit interval.

The uniform distribution in the interval [0, 1] is, for practical purposes, the only distribution that we need to generate for our simulations. The reason for this is that the samples from the other distributions are derived using the uniform distribution. The uniform distribution can be generated with either pseudorandom numbers or quasi-random numbers, and algorithms are available to transform a uniform distribution into any other distribution. The main and direct way to do this transformation is by cumulative distribution function inversion. To understand this method, let us consider the picture below, a cumulative distribution of a standard normal distribution that has a mean equal to zero and variance equal to 1, that is, N(0, 1).

3.7.6 The Moro's Inversion

A widely known general way to obtain the inverse of a normal distribution is by using the Box–Muller algorithm. However, for low-discrepancy sequences, it has been reported that this is not the best way because it damages low-discrepancy sequence properties. In addition, it has been reported that the Box–Muller algorithm is slower than Moro's inversion, which is described in what follows.

The traditional normal inversion algorithm is given by Beasley and Springer. However, this algorithm is not very good for tails of a normal distribution. The best way to obtain the inversion from U[0, 1] to a normal distribution is to use an algorithm presented in Moro (1995).

Moro modeled distribution tails using truncated Chebyshev series. Moro's algorithm divides the domain for Y (where Y = uniform sample) into two regions:

- The central region of the distribution, $|Y| \leq 0.42$, is modeled by the traditional algorithm.
- The tails of the distribution, $|Y| > 0.42$, are modeled with Chebyshev series.

Moro's algorithm is very simple. The only task is to write some constants and a few lines of code.

3.7.7 Implementation of Model

The magnetic properties of ultrafine and nanosized magnetic particles show significant variations from their bulk values. This has led to considerable interest at both the technological and fundamental levels. From the point of view of industrial applications, it is essential to obtain single-domain, magnetic fine-particle arrays with negligible size distribution of particles that are well separated and noninteracting. However, the standard techniques of synthesis result in clustering, a nonnegligible size distribution, and a variation in the shape of the particles. Recent experiments show that most of the magnetic properties, that is, the magnetization, coercivity, and Curie temperature, of these magnetic systems are very sensitive to the aforementioned parameters. A general description of the magnetization of fine magnetic particles is based on the theory of superparamagnetism, in which it is assumed that a magnetic particle is actually a single-domain entity in which the atomic magnetic moments rotate coherently such that the magnetization of the particle may be represented by a single magnetization vector with a large magnitude. The important feature of the system is its anisotropy. Uniaxial anisotropy, as often seen in transition metal oxides, leads to two equivalent magnetization states on each particle.

The relevant time scale in the system is the relaxation time, which is the time taken to reverse the magnetization of the particle from one equilibrium magnetization state to another. The relaxation time τ depends on the anisotropy constant K and the volume V of the particle as $\tau^{-1} \propto e^{-KV/K_BT}$, where T is the temperature of the system. The magnetic behavior observed in the system at different temperatures, however, depends both on the experimental time scale t_S for observations done and the relaxation time τ. For those temperatures where $\tau > t_S$, the system seems like a ferromagnet, and for temperatures at which $\tau < t_S$, the system seems superparamagnetic. The blocking temperature is determined by $\tau = t_S$ and separates the two regimes.

The magnitude of the magnetization on a single-domain magnetic nanoparticle is in itself a complicated problem that depends explicitly on the size and shape of the particle, and several studies of the magnetization properties of single-domain particles are available in the literature; these studies are based either on micromagnetic modeling or Monte Carlo simulations. From these simulations as well it was established that the magnetic properties of single-domain magnetic particles are strongly influenced by finite size and surface effects, such effects becoming more important as the size of the particles decreases.

For an array of single-domain magnetic nanoparticles numerical studies are further complicated by the inherent disorder in samples, as well as the interparticle interactions. The disorder in a system may be classified as (1) a spatial disorder in the actual positions of the grains in the array or (2) the possible randomness in the orientation of the easy axes of the different particles, depending on the shape and size of the particles. The relevant interactions are primarily the long-range dipolar interactions, and since in magnetic nanomaterials particles show a strong tendency to agglomerate, one must include the short-range exchange interactions. Numerical studies of the magnetic properties of single-domain magnetic arrays are based either on Langevin dynamics or Monte Carlo methods. Using these methods, hysteresis, magnetization, and magnetic susceptibility are calculated and simulated to study magnetization reversal and blocking temperatures. In the current work, we approached the matter using Monte Carlo methods. In our model, the hysteresis of an array of interacting single-domain magnetic particles is studied, where the particles interact via exchange and dipolar interactions.

The energy of a system is calculated according to the known formulation of the Hamiltonian for a system of interacting single-domain magnetic particles, each having a magnetic moment vector $\boldsymbol{\mu}_i$. The Hamiltonian is written as

$$E = -K\sum_i V_i \frac{(\boldsymbol{\mu}_i \cdot \mathbf{e}_i)^2}{|\boldsymbol{\mu}_i|^2} - \sum_{\{i \neq j\}} J_{ij}(\boldsymbol{\mu}_i \cdot \boldsymbol{\mu}_j) - \mu_0 \sum_{\{i \neq j\}} \frac{3(\boldsymbol{\mu}_i \cdot \mathbf{n}_{ij})(\boldsymbol{\mu}_i \cdot \mathbf{n}_{ij}) - \boldsymbol{\mu}_i \cdot \boldsymbol{\mu}_j}{r_{ij}^3} - \mu_0 \sum_j \mathbf{H} \cdot \boldsymbol{\mu}_j. \tag{3.27}$$

Here, the first term is the energy due to the anisotropy effect, with K as the anisotropy constant. The anisotropy energy associated with each particle depends on the volume of the particle V_i and the angle between its magnetic moment vector $\boldsymbol{\mu}_i$ and $\mathbf{e}_i$, which is the unit vector along the easy axis direction of the particle. The second term is the exchange energy, J_{ij} is the ferromagnetic exchange interaction between two particles with localized magnetic moment vectors $\boldsymbol{\mu}_i$ and $\boldsymbol{\mu}_j$, respectively, which for the purpose of simulation is assumed to have a site-independent constant value J. It is due to the Heisenberg effect between two touching nanoparticles (first-order neighbors), or nanoparticles that touch a third common nanoparticle (second-order neighbors). The third term is the dipolar interaction between the ith and jth particles, with $\mathbf{r}_{ij}$ as the distance between the particles; $\mathbf{n}_{ij}$ is the unit vector pointing along $\mathbf{r}_{ij}$ and is derived from the well-known Stoner–Wohlfarth model. Due to periodic boundary conditions applied to the system, the dipole–dipole interaction energy of the whole system (i.e., the bulk material) was calculated by means of a theoretical method for the summation of infinitely repeating lattice units. In this case, we exploited the Lekner summation. More details can be found in what follows and in the bibliography. The last term is the energy of the particles due to an externally applied magnetic field of magnitude H, i.e., the so-called energy due to the influence of the Zeeman effect.

It is assumed that the magnetic moment vector for a single particle has a temperature-independent constant value $\boldsymbol{\mu}_i = V_i M_S \boldsymbol{\sigma}_i$, where M_S is the saturation magnetization of the particle and $\boldsymbol{\sigma}_i$ with coordinates $(\sigma_i^x, \sigma_i^y, \sigma_i^z)$ is the unit vector along the direction of the magnetization vector. Several estimates of the actual value of M_S and the anisotropy constant K for a variety of nanomagnetic materials can be found in the literature. It is also possible to make a reasonable estimate of the average size as well as the shape and size variation of the particles in an actual sample using various characterization methods such as, for example, TEM and SEM. However, there seem to be no definite estimates for the exchange interaction parameter J.

Disorder occurs in the system since the magnetic grains are positioned randomly in the array. Eventually, it is possible to introduce another source of disorder by a random distribution of shape and size of single-domain grains. Because of the shape and size anisotropy in the sample, there is a random distribution of the magnitude and direction of the localized magnetic moment and in the direction of the easy axis of magnetization on the different particles in the sample. The positional disorder and particle concentration in the sample affect the dipolar interaction between two magnetic particles, which depends explicitly on the distance between the particles. The Lekner summation has been exploited to calculate the long-range interaction term in the case of periodic boundary conditions, as was imposed in our 2D and 3D models. In fact, usually, introducing a cutoff does not give the correct results for long-range interactions, and using an infinite summation method is better when imposing periodic boundary conditions, as well as when considering quasi-2D systems (i.e., periodic boundary conditions along one dimension, finite extension h along the other dimension) or quasi-3D systems (2D + h systems). However, the Lekner summation method is not the only method used. It is also possible to employ the Ewald summation method. Anyway, the Lekner infinite summation is preferable due to its faster convergence rate and its minor error in reproducing real systems. In computer simulations, the dependence of the magnetic properties on the distribution of grain sizes, the density of the grains, the anisotropy energy, and the exchange interactions in the array is investigated through Monte Carlo simulations for $\gamma - Fe_2O_3$ nanoparticle systems. In this section, we propose some of the most relevant results obtained by computer simulations. In the following figures, results could be presented by varying $J_{\mathrm{eff}} = M^2 V_0^2 J$, particle volume V_i, and particle density, which is varied by varying the occupation probability and, consequently, the length L of the simulation box. If the radii of nanoparticles are all equal to each other, then $E_A = KV_0$. In Fig. 3.9 we show an example of a Monte Carlo trend simulation for a $\gamma - Fe_2O_3$ nanoparticle system.

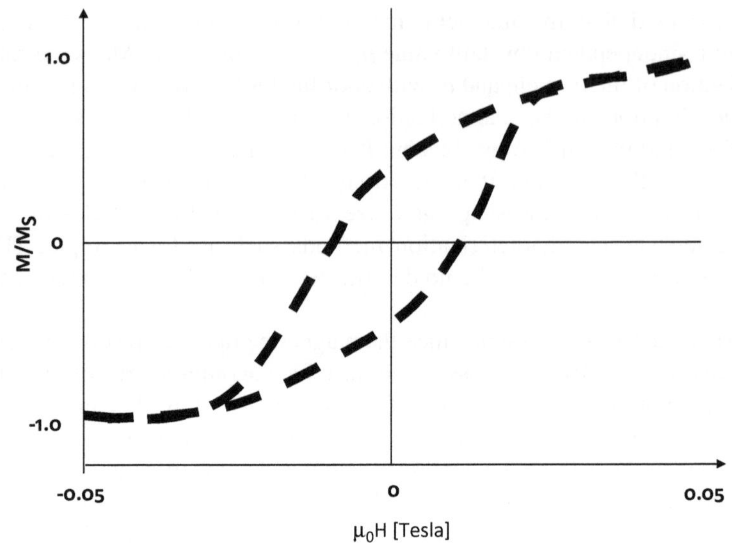

Fig. 3.9 Simulated hysteresis loops for $K = 0.046 \times 10^5$ J/m^3, fixed particle concentration, and uniform particle volume equivalent to a sphere of diameter 10 nm at different temperatures for $J_{eff} = 0$ at 5 K

3.8 Micro-/Nanosphere Modeling and Sensing

The modeling of spherical metallic nanoparticles is an important issue for the study of electromagnetic wave interactions and plasmonic resonances. It is possible to characterize the radiation behavior of a single gold nanoparticle (GNP) by varying the polarization source: the sensitivity of the radiation lobe direction will change with the light source polarization and by the GNP radius. Particularly interesting are techniques that make it possible to deposit single particles on a substrate, such as laser printing or a shape-controlled electrodeposition technique. The deposition and the localization of single GNPs represent an important issue regarding the light radiation process of nanosensors. In this framework, it is of considerable interest to investigate methods by which the radiation field spectra and local electric field can be focused and to study the radiation as a function of the nanostructure size of GNPs, of the source polarization. In particular, we characterize the electromagnetic radiation due to the main radiation lobes of a spherical GNP excited by a plane wave source. This characterization is useful for applying GNPs as antennas able to detect tissue anomalies or to couple the electromagnetic radiated field in a defined direction (useful aspect for implanted systems). The knowledge of the radiation behavior of a single GNP or aligned GNPs is important for biosensing applications where fluorescence emission and functionalizations are considered. Usually it is difficult to synthesize GNPs with constant dimensions, and a statistical study of the

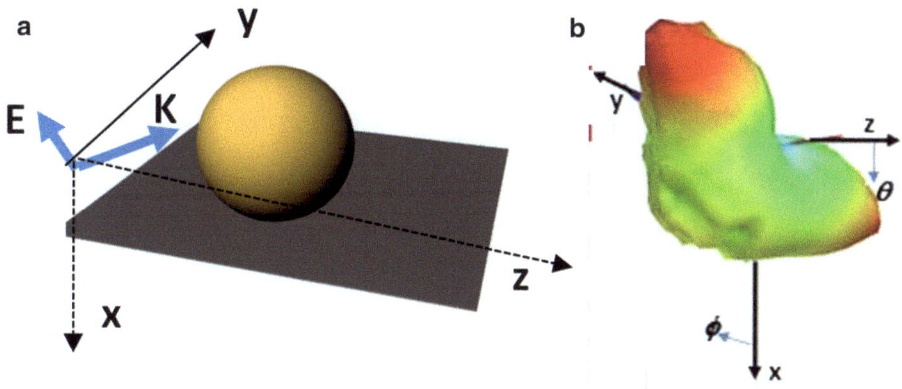

Fig. 3.10 (**a**) Modeling of a spherical GNP placed on metallic plane behaving as a ground plane and excited by a plane wave. (**b**) 3D radiation pattern for particular orientation of **K** wave vector

electromagnetic behavior of particles with different dimensions is necessary. A way to characterize the radiation properties of GNPs excited by a plane wave source is to establish the direction of light (in nanoshell configurations, two attached spheres radiate as a unique emitter). By aligning and coupling GNPs it is possible to define a precise radiation lobe for a fixed working wavelength and for a fixed light source polarization. The error of the directivity pattern due to the different dimensions of GNPs is reduced by considering attached and coupled GNPs that will stabilize the radiation lobe. GNPs can be modeled by a 3D FEM tool typically used for the design of micrometric structures. By assuming spherical particles, we characterized GNPs by means of a radiated field. This characterization allows us to study the electromagnetic behavior of micro-/nanoprobes using near-field detection systems. The FEM near-field region is the region closest to the plane wave source, and the electric field **V**(x, y, z), external to the region bounded by a closed spherical surface, may be written as

$$\mathbf{V}(x,y,z) = \int_S \left((j\omega\mu_0 \mathbf{I}_{\tan}) \frac{e^{-ik\left|\mathbf{r}-\mathbf{r}'\right|}}{|\mathbf{r}-\mathbf{r}'|} + \left(\mathbf{V}_{\mathbf{tan}} \times \nabla \frac{e^{-ik\left|\mathbf{r}-\mathbf{r}'\right|}}{|\mathbf{r}-\mathbf{r}'|} \right) + \left(\mathbf{V}_{normal} \nabla \frac{e^{-ik\left|\mathbf{r}-\mathbf{r}'\right|}}{|\mathbf{r}-\mathbf{r}'|} \right) \right) dS, \quad (3.28)$$

where S represents the radiation boundary surface, j is the imaginary unit, $\boldsymbol{I}_{\mathbf{tan}}$ is the component of the magnetic field that is tangential to the surface, $\boldsymbol{V}_{\mathbf{normal}}$ is the component of the electric field that is normal to the surface, k is the free space wave number, and $\boldsymbol{r}$ and $\boldsymbol{r}'$ represent field points and source points on the surface, respectively. The value of a vector field quantity, such as the ***I***-field (magnetic field) or ***I***-field (electric field), at points inside each tetrahedron

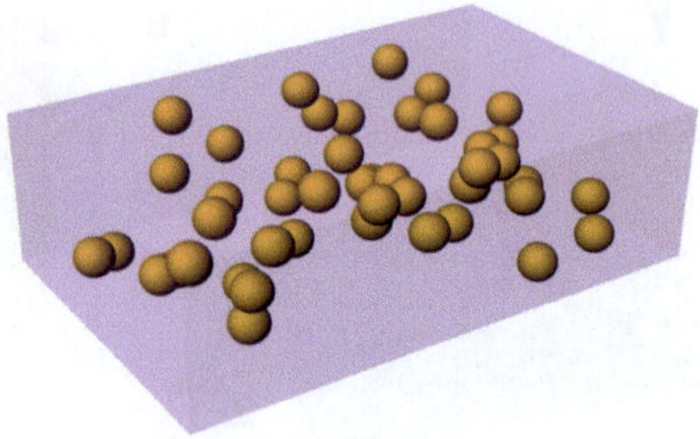

Fig. 3.11 Example of spherical GNPs embedded in polymer

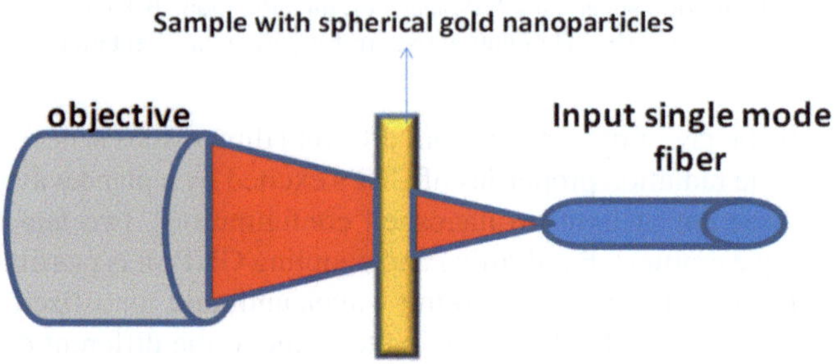

Fig. 3.12 Scheme of experimental setup for measure of transmittivity

is interpolated from the vertices of the tetrahedrons that discretize the system. Figure 3.10a, b illustrates the 3D FEM modeling and the total electric field distribution of a spherical GNP placed in the equivalent metallic surface (thin film ground plane boundary condition): the GNP is excited by a plane wave characterized by an electric field ***E*** orthogonal to the wave vector ***K*** (definition of plane wave). To define a traveling plane wave source, we introduce a rectangular ground plane guiding a defined polarized light. Other applications involving spherical nanoparticles include biomedical imaging by fluorescence emission, light trapping, nanosensors for robotics, mechanical reinforcement, and drug delivery. The introduction of spherical nanoparticles could change the physical properties of a polymer such as, for example, piezoelectricity, electrical conductivity, light absorption, and viscoeleasticity. Spherical GNPs can be generated into a polymer by proper chemical processes (such as chemical reduction using a gold precursor). And nowadays, the challenge is to control the monodispersion of nanofillers to improve light enhancement and a high quality (Q) factor, as happens for uniform gratings. Figure 3.11 shows a scheme of polydispersion of spherical GNPs generated in a polymeric matrix. To measure the transmittivity of the nanocomposite sample, we can use the simple setup of Fig. 3.12.

3.9 Photonic Crystal Sensors: An Overview

Photonic crystals are used in many optical sensing applications. There are many configurations and the main layouts used are the square lattice and triangular lattice. A particular layout that can improve high resonant modes such as the Whispering Gallery Modes (WGM) is the circular layouts. Typically they are composed by air holes in a semiconductor material, but there are also pillar-type photonic crystals that guarantee a high dielectric contrast. The nanometric pillars are growth on a substrate and can be implemented for plasmonic applications ted using metallic materials. Photonic crystals can be 1D, 2D, or 3D if the layout periodicity is in a one-dimensional, two-dimensional, or three-dimensional plane, respectively. To improve a high localized electromagnetic energy, microcavities are fabricated by missing a central hole or a central pillar. To optimize the energy localized in a microcavity and to increase the quality factor ($Q = \Delta\lambda/\lambda_0$), holes near the cavity are shifted or their diameters are changed. The quality factors are defined in the *inplane* region ($Q\ //$), defining the energy resonating in the layout plane, and in the vertical plane ($Q\ \perp$), defining the energy in the vertical plane and emitted by the microcavity. To measure the light emitted in the vertical plane, a micro photoluminescence setup can be used. In this experimental setup, the visible light is reflected by a dichroic mirror and focalized by a microscope objective over the sample, thereby obtaining an excitation micrometric spot. The infrared light emitted from the microcavity is collected in backscattering by the same microscope objective and then transmitted by the dichroic mirror and by a core fiber to an optical analyzer. The microcavity can also be excited by bending a fiber, as discussed in the previous chapter. With regard to the *inplane* light coupling, the microcavity can be excited by a $W1$ defect line coupled to input and output tapered ridge waveguides. To increase the vertical quality factor, it is possible to improve the technology by implementing membrane-type photonic crystals that have quantum dots. The air membrane separates the photonic crystal slab from the substrate supporting the air cavity resonance: the air cavity depth should be on the order of $m\lambda_0/2$, with $m = 1,3,5,\ldots$ (the best cavity dimension is considered by observing the substrate loss). The quantum dot emitters (emitting around λ_0) are chosen to increase the light radiated by the microcavity and are placed in different layers of the photonic crystal slab.

Implementing properly designed electrodes, it is possible to realize photonic crystal lasers. Photonic crystals are usually simulated by FDTD, FEM, or by a wave expansion method (WEM). The WEM is useful for defining the guided modes depending on the crystal directions. To improve the maximum energy transferred by the photonic crystal slab, an accurate design takes into account the single-mode condition. The steps to follow to design a cavity with a confined are as follows:

- Define a band gap inside the light cone for a photonic crystal slab without a central microcavity.
- Separate the band into two bands to localize a single mode between them.
- Optimize the geometrical parameters to improve the maximum quality factor.

The applications are different and involve frequency filtering, on/off power switches, telecommunications systems (wavelength division multiplexing), or biosensing (such as DNA lab on chip or biotargeting in general).

Acknowledgments The author gratefully acknowledges the help of Dr. Matteo Cacciola.

References

Chib, S., Greenberg, E.: Understanding the Metropolis–Hastings algorithm. Am. Stat. **49**(4), 327–335 (1995)

Moro, B.: The full Monte. Risk Mag. **8**(2), 57–58 (1995)

Simon, H.J., Mitchell, D.E., Watson, J.G.: Surface plasmons in silver films a novel undergraduate experiment. Am. J. Phys. **43**(7), 630–636 (1975)

Venugopal, R.S., Moutzouris, K., Ebrahimzadeh, M.: Nonlinear frequency con-version in semiconductor optical waveguides using birefringent, modal and quasi-phase-matching technique. J. Opt. A: Pure Appl. Opt. **6**, 73241–73249 (2004)

Chapter 4
Small Antennas, Wireless System and Small Sensors

Abstract This chapter discusses piezoelectric microelectromechanical system (MEMS) circuits and small antennas. In particular, small antennas dedicated to sensing are discussed using a theoretical approach explaining the physical aspects. The goal of this chapter is to provide tools for the analysis of micro-/nanoprobes starting with an electromagnetic characterization. In addition, some applications are proposed.

4.1 Characterization of Small Piezoelectric Devices

Piezoelectric material converts energy between the mechanical and electrical domains. The development of energy-harvesting devices for different energy sources has been the subject of intense study. In particular, mechanical vibrations are easily exploited through microelectromechanical system (MEMS) technology. Mechanical vibration energy can be converted into electric energy by piezoelectric transducers, which are preferable to electromagnetic and electrostatic transducers due to their higher energy density. Piezoelectric mechanical transducers are based on the mechanical deformations that occur due to a bimorphic structure (piezoelectric/structural layer) induced by the internal stress generated in the piezoelectric material. Among possible piezoelectric thin films, $Pb(Zr,Ti)O_3$ (PZT), ZnO, and AlN (aluminum nitride) are the most widely used in piezoelectrically actuated mass sensors. Despite its lower piezoelectric coefficient, AlN is very attractive for micromechanical energy harvesting due to the simple manufacturing process and easy integration with complementary metal–oxide semiconductors (CMOSs). The technology makes it possible to control the stress/strain properties of a cantilever beam by controlling the thickness. Particular kinds of basic MEMS are made by a piezoelectric layer embedded in a metallic layer, as illustrated in Fig. 4.1a, b, assuming a cantilever layout. The mechanical deformation of the piezoelectric device leads to an electrical signal using the piezoelectric effect: an external

A. Lay-Ekuakille, *Optical Waveguiding and Applied Photonics*, Lecture Notes in Nanoscale Science and Technology 10, DOI 10.1007/978-1-4614-5959-0_4,

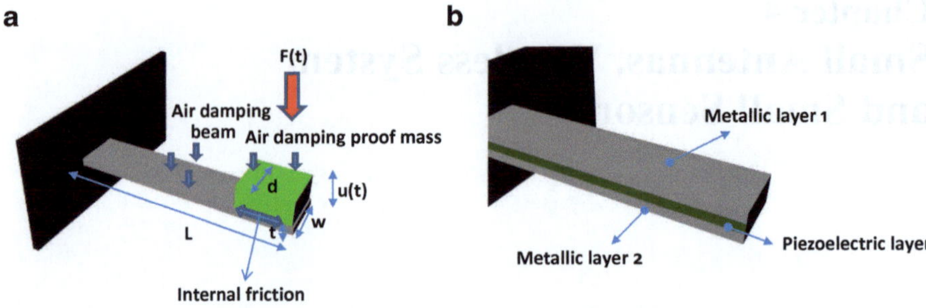

Fig. 4.1 (**a**) Losses of piezoelectric cantilever element: damping mechanisms and internal friction of a cantilever in air. (**b**) Complete cantilever piezoelectric structure

force $F(t)$ shakes the cantilever beam in the (r,θ) plane (this plane is also useful for the characterization of circular layouts); the vibration, indicated by the function $u(t)$, will generate through the piezoelectric effect a voltage V_p/current i signal at the output of the system. The system must be matched by an external load Z to transfer the maximum energy. The cantilever beam can be modeled through the following relationships between the stress **K**, strain **P**, electric field **V**, and the electric displacement field **E**:

$$\begin{aligned} \mathbf{K} &= c\mathbf{P} - e^T\mathbf{E}, \\ \mathbf{E} &= e\mathbf{P} - \varepsilon\mathbf{V}, \end{aligned} \tag{4.1}$$

where c, e, and ε are the elasticity matrix, the coupling matrix, and the relative permittivity matrix, respectively. The displacement is a solution of the following electromechanical system of equations:

$$\begin{aligned} i(t) &= b\ u(t) - C_f\ i(t), \\ Ku(t) &= F(t) - bi(t) - \rho\ u(t) + m_{eff}\ddot{u}\,(t), \end{aligned} \tag{4.2}$$

where i(t) is the current generated by the vibration flowing in the load, b is a geometric coefficient linked with the piezoelectric coefficient of material, $A = w{\cdot}L$ is the cantilever surface, C_f is the free capacitance (C/V), K is the stiffness, m_{eff} is the effective mass of the system, $F(t) = F_0\sin(\omega t)$ is the time domain sinusoidal mechanical excitation (with ω the angular frequency of vibration and F_0 the constant magnitude), ρ is the effective damping factor, which includes all the losses reported in Fig. 4.1a (internal friction, air damping beam, and air damping proof mass). We note that the initial conditions [of equation system (4.2)] $u(t = 0) = 0$, $i(t = 0) = 0$, and $\ddot{u}(t = 0) = 0$ correspond to a non vibrating cantilever beam at the time $t = 0$. We note that for maximum vibrating efficiency, $F(t)$ must be in phase with $u(t)$ and have the same angular frequency of the natural cantilever's resonance. Therefore, to fix the working frequency, it is necessary to calculate the modes (resonance conditions) of the cantilever by means of the following dispersion equation:

$$1 + \cos(k_n L)\cosh(k_n L) = 0. \tag{4.3}$$

The design criteria for an efficient energy-harvesting cantilever beam are developed in the points listed below:

(i) Setting of external vibrating force F(t): for a maximum cantilever displacement and, consequently, for maximum generated power due to the piezoelectric effect, the external force must have the frequency of the resonant mode; ***(ii) use of a proof mass*** m_{eff}: the adding of proof masses allows one to operate at lower frequencies compared with the natural frequencies of the uniform cantilever beam. The addition of m_{eff} will decrease the losses due to the internal friction and to the damping losses; ***(iii) choice of geometrical parameters***: the choice of the geometrical parameters is important to decrease the losses. In any case, the choice of the parameters must be in agreement with the technological aspects; ***(iv) choice of optimal output load (energy-harvesting optimization)***: to transfer a lot of generated power, a matched load must be considered at the output of the system. The load must guarantee higher electromechanical coupling efficiency.

A good design is possible by increasing the effective mass (especially by adding a proof mass) and by decreasing the working vibration frequency.

Using AlN as the piezoelectric material sandwiched between two molybdenum (Mo) electrodes it is possible to implement an efficient piezoelectric cantilever beam. The AlN thin film could be grown on a substrate consisting of a thin SiO_2 sacrificial layer and growth by a radio frequency (RF) sputtering technique. A complete energy-harvesting experimental setup based on vibrational forces should consider a controlled shaker and by a lock-in-amplifier (Massaro et al. 2011) and a matched load at the output of the energy harvesting device. In addition, the maximum displacement of the piezoelectric cantilever due to an applied voltage could be measured by an image postprocessing method using a camera coupled to a microscope. The error of the postprocessing image analysis is due to the error of the image reading of the software being used.

4.1.1 Example Piezoelectric Antennas by a Freestanding Structure

By acting on the technology it is possible to fabricate a freestanding structure characterized by a curvature (Fig. 4.2). This curvature could be used to transmit a signal as a probe antenna. The possible geometrical deformation of the freestanding structure (due to frequencies of resonance) allows us to use two freestanding structures as a transmitter /receiver radiation system. The deformation can be due to the application of a voltage signal, as illustrated by Fig. 4.3a. We observe that a high ΔV

Fig. 4.2 Freestanding piezoelectric structure behaving as a small antenna

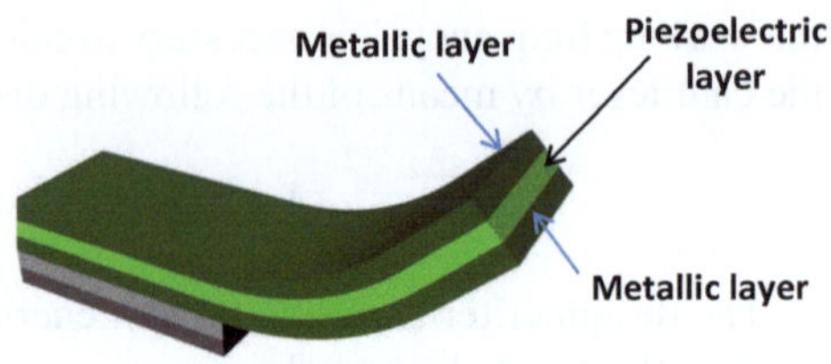

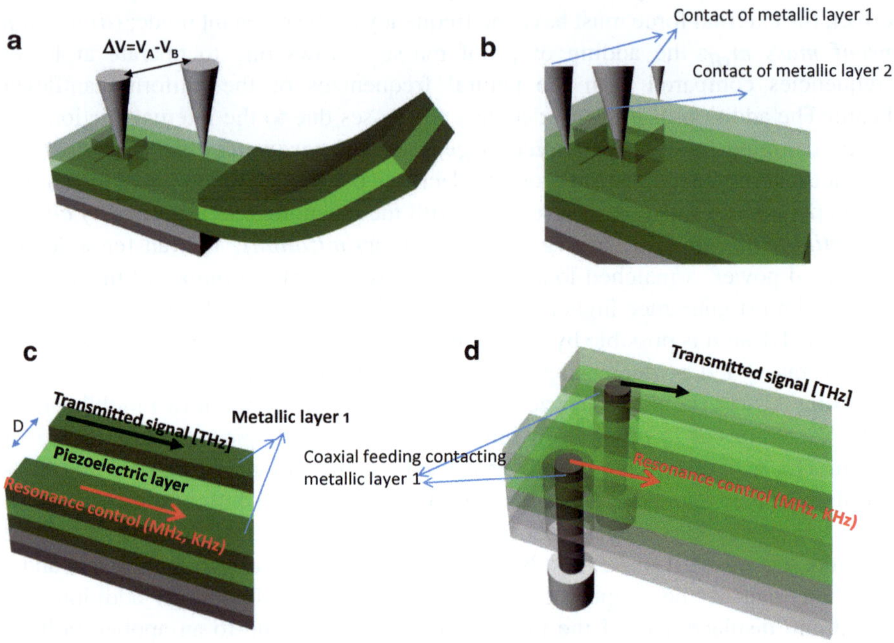

Fig. 4.3 (**a**) Electrical contact of LCR meter providing a voltage. (**b**) Cross section of contact. (**c**) *Top view* of a possible implementation concerning manufacturing. (**d**) Cross section of coaxial feeding configuration

voltage amplitude could destroy and "burn" the piezoelectric antenna. For this reason, it necessary to study before the effect of the input power to define the limits and the quality of the device. The geometrical deformation of a transmitter ring can modulate the phase or the amplitude (variation of the radiation resistance due to geometrical deformation) of a transmitted kilohertz/terahertz signal: the modulation is due to the piezoelectric effect of the piezoelectric layer and is performed by applying a resonance control signal, which can be found at lower frequencies.

Resonant frequencies can be measured by an LCR meter instrument that can also characterize the equivalent capacitance, inductance, and resistance of the whole system. Moreover, the LCR meter will provide the total impedance phase θ ($Z = Xe^{j\theta}$), which is a useful parameter for matching impedance processes. As observed in Fig. 4.3b, to excite the small antenna, one microtip should contact metallic layer 2, and the other two microtips should contact metallic layer 1. Both the kilohertz signal (a typical signal related to the resonance of a piezoelectric layer) and the terahertz signal (related to the radiated signal) should be applied on the same metallic conductive layer 1 by means of the configuration reported in the scheme of Fig. 4.3c, where the distance D allows one to ignore the unwanted interaction between the two signals. Both the kilohertz/terahertz signals can also be coupled to the freestanding structure by the coaxial feeding sources reported in Fig. 4.3d.

Chemical processes based on lithography and wet chemical etching processes can be applied to perform excitation layouts and antenna layouts. A better radiation behavior of freestanding cantilever can be performed by improving the curvature during the fabrication process (control of the curvature can be carried out by tuning the piezoelectric and the metallic layer thicknesses during the RF sputtering deposition). The curvature will enhance the deformation of the structure at piezoelectric frequencies and, consequently, optimize the modulation of the signal. In this case, the principle of amplitude modulation for a transmitter ring is based on the piezoelectric resonance, which changes the radius of curvature, and, consequently, the radiated power $Pr = Rr|I_0|^2$, where I_0 is the current traveling the radiation resistance Rr. Simulating a bent AlN antenna with a radius of 50 μm (assuming a perfect ring layout), we found a working frequency of around 2 THz. We observe that an antenna can be modeled on the assumption that the metallic layers are the metallic boundary conditions. The working frequency and the radiation pattern will mainly be functions of the radius of curvature and of the piezoelectric thickness. The small piezoelectric antennas could be implanted for human body activities considering a biocompatible packaging.

4.2 FDTD-Based Sensor Modeling

The design and full-wave analysis of complex sensing devices require accurate electromagnetic field prediction models (Mukhopadhyay 2005; Chen and Chew 2003; Caratelli and Yarovoy 2010). One widely used technique is the finite-difference time-domain (FDTD) algorithm (Yee 1966; Taflove and Hagness 2005). The FDTD method belongs in the general class of mesh-based numerical techniques for solving the time-dependent Maxwell's equations. The time-domain formulation of the method is an attractive feature in various contexts. In particular, it should be noted that, where a broadband pulse is used as the source, the electromagnetic response of the structure under analysis over a wide range of frequencies can be derived by running a single numerical simulation, whereas frequency-domain algorithms, such as the finite-element method (FEM), require the response of the device to be computed for each frequency point (Jin 2002). This is very useful in applications where the resonant frequencies of the structure of interest are not known a priori or whenever a broadband analysis is needed.

4.2.1 Subcell Averaging FDTD Scheme

In the conventional formulation of the FDTD algorithm (Yee 1966; Taflove and Hagness 2005), each cell in the computational grid is implicitly supposed to be filled by a homogeneous material. For this reason, the adoption of Cartesian meshes could result in reduced numerical accuracy where structures having complex geometry are to be modeled. In this context, the locally conformal FDTD scheme detailed subsequently provides a clear advantage over a staircasing approach or unstructured and stretched space lattices, which would potentially suffer from significant numerical dispersion or instability (Taflove and Hagness 2005). Such a scheme, featuring an enhanced numerical accuracy compared to a conventional algorithm, is based on the definition of effective material tensors accounting for the local electrical and geometrical properties of the device under analysis (Caratelli and Cicchetti 2003).

Let us consider a 3D domain $\mathcal{D}$ filled by a linear, isotropic, nondispersive material, with permittivity $\varepsilon(\mathbf{r})$, magnetic permeability $\mu(\mathbf{r})$, and electrical conductivity $\sigma(\mathbf{r})$. In such a domain, a dual-space, nonuniform lattice formed by a primary and secondary Cartesian mesh is introduced. The primary mesh $\mathcal{M}_{\mathcal{D}}$ is composed of space-filling hexahedrons having nodes $\mathbf{r}_{i,j,k}$. Thus, upon denoting the triplet $(i, j, k) \in \mathbb{N}_0^3$ as $\mathbf{\Upsilon}$ for shortness, the edge lengths between adjacent vertices in $\mathcal{M}_{\mathcal{D}}$ turn out to be $\Delta\xi_{\Upsilon} = \hat{\boldsymbol{\xi}} \cdot \left(\mathbf{r}_{\Upsilon+\Lambda_\xi} - \mathbf{r}_{\Upsilon}\right)$, where $\mathbf{\Lambda}_\xi$ and $\hat{\boldsymbol{\xi}}$ are the fundamental triplet and the unit vector, respectively, which are relevant to the ξ coordinate axis ($\xi = x, y, z$). On the other hand, the secondary or dual mesh $\widetilde{\mathcal{M}}_{\mathcal{D}}$ consists of closed hexahedrons whose edges penetrate the shared faces of the primary cells and connect the relevant centroids $\widetilde{\mathbf{r}}_{\Upsilon}$. As a consequence, the dual edge lengths are found to be $\xi_{\Upsilon} = \hat{\boldsymbol{\xi}} \cdot \left(\widetilde{\mathbf{r}}_{\Upsilon} - \widetilde{\mathbf{r}}_{\Upsilon-\Lambda_\xi}\right) = \left(\Delta\xi_{\Upsilon-\Lambda_\xi} + \Delta\xi_{\Upsilon}\right)/2$. As usual, the electric field components are defined along each edge of a primary lattice cell, whereas the magnetic field components are assumed to be sampled at midpoints of the edges of the secondary lattice cells.

The governing relationship between the E - and H - field components is given by Maxwell's equations in integral form, that is, the Faraday–Neumann and Ampere laws (Stratton 2007). In particular, the enforcement of Ampere's law on the general dual-mesh cell face $\widetilde{S}_\xi\big|_{\Upsilon}$, having boundary $\partial\widetilde{S}_\xi\big|_{\Upsilon} = \widetilde{C}_\xi\big|_{\Upsilon}$, results in the integral equation

$$\oint_{\widetilde{C}_\xi|_{\Upsilon}} \mathbf{H}(\mathbf{r},t) \cdot d \quad = \iint_{\widetilde{S}_\xi|_{\Upsilon}} \sigma(\mathbf{r}) E_\xi(\mathbf{r},t) dS + \frac{\partial}{\partial t} \iint_{\widetilde{S}_\xi|_{\Upsilon}} \varepsilon(\mathbf{r}) E_\xi(\mathbf{r},t) dS. \tag{4.4}$$

Under the assumption that the spatial increments $\Delta\,\xi_{\Upsilon}$ of the computational grid are small compared to the minimum operating wavelength, the contour integral appearing on the left-hand side of (4.4) can be evaluated using the mean value theorem, neglecting infinitesimal terms of higher order. In addition, the time derivative in (4.4) can be approximated using a central-difference approximation, which is second-order accurate where the E – and H – field components are staggered in the time domain. Therefore, upon introducing the normalized field quantities

$$\mathcal{E}_\xi\big|_{\Upsilon}^{n} = \Delta\xi_{\Upsilon} E_\xi\left(\mathbf{r}_{\Upsilon} + \frac{1}{2}\hat{\boldsymbol{\xi}}\,\Delta\xi_{\Upsilon}, t_n\right), \tag{4.5}$$

$$\mathcal{H}_\xi\big|_{\Upsilon}^{n+\frac{1}{2}} = \quad {}_{\xi_{\Upsilon}} H_\xi\left(\widetilde{\mathbf{r}}_{\Upsilon} - \frac{1}{2}\hat{\boldsymbol{\xi}}\,\Delta\xi_{\Upsilon}, t_{n+\frac{1}{2}}\right), \tag{4.6}$$

with $t_\nu = \nu\Delta t$, where Δt is the time step selected according to the Courant–Friedrichs–Lewy (CFL) stability condition (Taflove and Hagness 2005), after some algebraic manipulation the following explicit time-stepping equation is obtained:

$$\mathcal{E}_\xi\big|_{\Upsilon}^{n+1} = \frac{\overline{\varepsilon}_\xi\big|_{\Upsilon} - \frac{1}{2}\overline{\sigma}_\xi\big|_{\Upsilon}\Delta t}{\overline{\varepsilon}_\xi\big|_{\Upsilon} + \frac{1}{2}\overline{\sigma}_\xi\big|_{\Upsilon}\Delta t}\mathcal{E}_\xi\big|_{\Upsilon}^{n} + \frac{\Delta t}{\overline{\varepsilon}_\xi\big|_{\Upsilon} + \frac{1}{2}\overline{\sigma}_\xi\big|_{\Upsilon}\Delta t}(\Delta \times \mathcal{H})_\xi\Big|_{\Upsilon}^{n+\frac{1}{2}}, \tag{4.7}$$

where $(\nabla \times \mathcal{H})_\xi\Big|_{\Upsilon}^{n+\frac{1}{2}} = \mathcal{H}_\zeta\big|_{\Upsilon}^{n+\frac{1}{2}} - \mathcal{H}_\zeta\big|_{\Upsilon-\Lambda_\eta}^{n+\frac{1}{2}} - \mathcal{H}_\eta\big|_{\Upsilon}^{n+\frac{1}{2}} + \mathcal{H}_\eta\big|_{\Upsilon-\Lambda_\zeta}^{n+\frac{1}{2}}$ denotes the finite-difference expression of the ξ - component of the normalized magnetic field curl. By applying the duality principle in the discrete space $\left(\mathcal{M}_{\mathcal{D}}, \widetilde{\mathcal{M}}_{\mathcal{D}}\right)$, the FDTD update equations of the H - field can be easily derived as

$$\mathcal{H}_\xi\big|_{\Upsilon}^{n+\frac{1}{2}} = \mathcal{H}_\xi\big|_{\Upsilon}^{n-\frac{1}{2}} - \frac{\Delta t}{\overline{\mu}_\xi\big|_{\Upsilon}}(\nabla \times \mathcal{E})_\xi\Big|_{\Upsilon}^{n}, \tag{4.8}$$

with $S_\xi|_{\Upsilon}$ being the general primary-mesh cell face orthogonal to the ξ -coordinate axis. In (4.7) and (4.8), the information regarding the local physical properties, as well as the geometrical nonconformability of the electromagnetic structure under analysis to the adopted Cartesian mesh, is transferred to the following position-dependent effective permittivity, electrical conductivity, and permeability tensors:

$$\left\{\begin{matrix}\overline{\varepsilon}\\ \overline{\sigma}\end{matrix}\right\}\Bigg|_{\xi}\Bigg|_{\Upsilon} = \frac{1}{\Delta\xi_{\Upsilon}}\iint_{\widetilde{S}_\xi|_{\Upsilon}}\left\{\begin{matrix}\varepsilon\\ \sigma\end{matrix}\right\}(\mathbf{r})dS, \tag{4.9}$$

$$\overline{\mu}_\xi\big|_{\Upsilon} = \frac{1}{\xi_{\Upsilon}}\iint_{S_\xi|_{\Upsilon}}\mu(\mathbf{r})dS. \tag{4.10}$$

In this way, a significant enhancement in terms of numerical accuracy can be achieved over the conventional staircase modeling approach. It is worth noting that the computation of the effective material parameters [(4.9) and (4.10)] can be conveniently carried out before the FDTD-method time marching starts. As a consequence, unlike in conformal techniques based on stretched computational grids, no additional correction is required in the core of the numerical algorithm.

4.2.2 Near-Field to Far-Field Transformation

The FDTD method is inherently a near-field technique that allows for accurately computing the electromagnetic field distribution within a volume surrounding the structure under analysis. To perform the field computation outside the FDTD domain, a frequency-domain model based on the surface equivalence theorem (Stutzman and Thiele 1997) and the free-space dyadic Green's functions (DGFs) (Felsen and Marcuvitz 1994) can be conveniently employed. This approach has an intrinsic advantage with respect to the usual techniques based on the vector potentials, or Kirchhoff's surface integral representation (KSIR) (Ramahi 1997), since it does not entail any numerical derivative operation for evaluating the field quantities of interest. An additional advantage of the mentioned near-field to far-field (NF-FF) transformation consists in the ability to distinguish among the radiative, inductive, and electrostatic field components (Felsen and Marcuvitz 1994) and, consequently, to investigate the corresponding spatial transition regions of the electromagnetic field. This analysis is in turn important because it provides a useful insight into the underlying physical mechanisms that are responsible for the behavior of the structure (Fig. 4.4).

Field computation based on the aforementioned approach requires the evaluation of the time-domain electric and magnetic equivalent currents $\mathbf{J}_S(\mathbf{r},t)$ and $\mathbf{M}_S(\mathbf{r},t)$ along the Huygens surface S enclosing the device under analysis (Fig. 1). Then an on-the-fly Fourier transformation, in step with the FDTD numerical simulation, is performed using the following running-sum implementation of the DFT:

$$\widetilde{\psi}^{(n)}(\mathbf{r},f) = \widetilde{\psi}^{(n-1)}(\mathbf{r},f) + \psi|_{\mathbf{r}}^{n} e^{-j2\pi f\, n\,\Delta t}, \tag{4.11}$$

where

$$\widetilde{\psi}^{(n-1)}(\mathbf{r},f) = \sum_{\nu=1}^{n-1} \psi|_{\mathbf{r}}^{\nu} e^{-j2\pi f\, \nu\,\Delta t}, \tag{4.12}$$

with ψ denoting the general current component. Finally, the electromagnetic field distribution excited outside S is determined by numerical evaluation of the convolutional integral between the free-space DGFs and the frequency-domain equivalent currents (Felsen and Marcuvitz 1994) as follows:

$$\mathbf{E}(\mathbf{r},f) = \sum\nolimits_E^{(x)}(\mathbf{r},f) + \sum\nolimits_E^{(y)}(\mathbf{r},f) + \sum\nolimits_E^{(z)}(\mathbf{r},f), \tag{4.13}$$

with

$$\begin{aligned} \sum\nolimits_E^{(x)}(\mathbf{r},f) = \sum_{p=1,2}\sum_{j}\sum_{k}\Big[{}^{E}\underline{\mathbf{G}}^{J}\Big(\mathbf{r}\Big| x_{i_{S_p}}, y_{j+\frac{1}{2}}, z_{k+\frac{1}{2}}, f\Big) \cdot \mathbf{J}_S|^{f}_{i_{S_p}, j+\frac{1}{2}, k+\frac{1}{2}} \\ + {}^{E}\underline{\mathbf{G}}^{M}\Big(\mathbf{r}\Big| x_{i_{S_p}}, y_{j+\frac{1}{2}}, z_{k+\frac{1}{2}}, f\Big) \cdot \mathbf{M}_S|^{f}_{i_{S_p}, j+\frac{1}{2}, k+\frac{1}{2}}\Big] \Delta y_j \Delta z_k, \end{aligned} \tag{4.14}$$

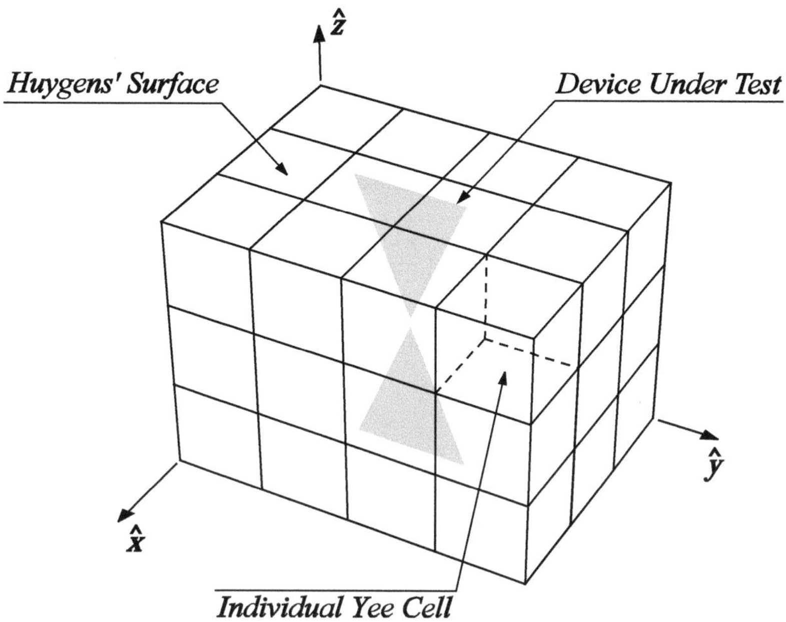

Fig. 4.4 Huygens surface enclosing an electromagnetic structure

$$\Sigma_E^{(y)}(\mathbf{r},f) = \sum_{p=1,2}\sum_i\sum_k \left[{}^E\underline{\mathbf{G}}^J\left(\mathbf{r}\middle|x_{i+\frac{1}{2}}, y_{j_{S_p}}, z_{k+\frac{1}{2}}, f\right) \cdot \mathbf{J}_S|^f_{i+\frac{1}{2}, j_{S_p}, k+\frac{1}{2}} + {}^E\underline{\mathbf{G}}^M\left(\mathbf{r}\middle|x_{i+\frac{1}{2}}, y_{j_{S_p}}, z_{k+\frac{1}{2}}, f\right) \cdot \mathbf{M}_S|^f_{i+\frac{1}{2}, j_{S_p}, k+\frac{1}{2}} \right] \Delta x_i \Delta z_k, \tag{4.15}$$

$$\Sigma_E^{(z)}(\mathbf{r},f) = \sum_{p=1,2}\sum_i\sum_j \left[{}^E\underline{\mathbf{G}}^J\left(\mathbf{r}\middle|x_{i+\frac{1}{2}}, y_{j+\frac{1}{2}}, z_{k_{S_p}}, f\right) \cdot \mathbf{J}_S|^f_{i+\frac{1}{2}, j+\frac{1}{2}, k_{S_p}} + {}^E\underline{\mathbf{G}}^M\left(\mathbf{r}\middle|x_{i+\frac{1}{2}}, y_{j+\frac{1}{2}}, z_{k_{S_p}}, f\right) \cdot \mathbf{M}_S|^f_{i+\frac{1}{2}, j+\frac{1}{2}, k_{S_p}} \right] \Delta x_i \Delta y_j, \tag{4.16}$$

expressing the electric field contributions due to the currents flowing along the faces of the Huygens surface that are orthogonal to the $x-$, $y-$, and $z-$coordinate directions, respectively. Dual equations hold for the magnetic field $\mathbf{H}(\mathbf{r}, f)$.

4.2.3 Electromagnetic Characterization of Miniaturized Sensing Devices

Using the aforementioned subcell method, the full-wave analysis of a miniaturized electromagnetic sensor, modeled as a dipole operating in free space, was carried out. In particular, the arm's length, radius, and feeding delta gap of the dipole were selected to be $l_d = 0.4385$ mm, $r_d = 0.01$ mm, and $\delta_d = 0.02$ mm, respectively, in such a way as to achieve a fundamental resonant frequency of $f_c \simeq 150$ GHz. In doing so, the individual device was meshed on a graded space lattice with the maximum spatial increment $\Delta\ \xi_{\max} = \lambda_{\min}/20$, where $\lambda_{\min}$ denotes the wavelength in free space at the maximum operating frequency relevant to the spectrum of the excitation signal, which is a Gaussian-modulated sinusoidal voltage pulse defined by

$$\Pi_g(t) = V_g \exp\left[-\left(\frac{t - t_0}{T_g}\right)^2\right] \sin(2\pi f_c t) u(t), \tag{4.17}$$

where $u(t)$ is the usual unit-step distribution, and

$$\frac{t_0}{4} = T_g = \frac{2\sqrt{\ln 10}}{\pi B}. \tag{4.18}$$

The selection of the time parameter T_g according to (4.15) results in a significant spectral content of the source pulse, measured at -10 dB level, within a frequency range of $f_{\min} = f_c - B/2$ to $f_{\max} = f_c + B/2$, with a bandwidth of $B = 100$ GHz. The energy delivered by the voltage generator, assumed to feature internal resistance $R_g = 50\ \Omega$, is accounted for by adding a local current density term in the finite-difference equations used to update the transient distribution of the electric field at the feed section of the sensor probe:

$$\mathcal{E}_\xi\Big|_{\Upsilon_g}^{n+1} = \frac{\overline{\varepsilon}_\xi\Big|_{\Upsilon_g} - \frac{\Delta t}{2R_g}}{\overline{\varepsilon}_\xi\Big|_{\Upsilon_g} + \frac{\Delta t}{2R_g}} \mathcal{E}_\xi\Big|_{\Upsilon_g}^{n} + \frac{\Delta t}{\overline{\varepsilon}_\xi\Big|_{\Upsilon_g} + \frac{\Delta t}{2R_g}} \left[(\Delta \times \mathcal{H})_\xi\Big|_{\Upsilon_g}^{n+\frac{1}{2}} + \mathcal{I}_g\Big|^{n+\frac{1}{2}}\right], \tag{4.19}$$

where Υ_g is the index triplet relevant to the driving point, and $\mathcal{I}_g\Big|^{n+\frac{1}{2}} = \Pi_g\left(t_{n+\frac{1}{2}}\right)/R_g$ denotes the discrete-valued excitation current. In this way, the voltage and current accepted at the input terminals of the feeding line can be computed as follows:

$$v_{in}(t_n) = -\int_{C_V} \mathbf{E}(\mathbf{r}, t_n) \cdot d \quad \simeq -\mathcal{E}_\xi\Big|_{\Upsilon_g}^{n}, \tag{4.20}$$

$$i_{in}\left(t_{n+\frac{1}{2}}\right) = \oint_{C_I} \mathbf{H}\left(\mathbf{r}, t_{n+\frac{1}{2}}\right) \cdot d \quad \simeq (\nabla \times \mathcal{H})_\xi\Big|_{\Upsilon_g}^{n+\frac{1}{2}}, \tag{4.21}$$

where C_V is an open contour extending along the driving point, and C_I is a closed contour path wrapping around the generator. Then the input impedance of the device is given by

$$Z_{in}(f) = e^{-j\pi f \Delta t} \frac{\mathcal{DFT}[v_{in}(t_n)](f)}{\mathcal{DFT}\left[i_{in}\left(t_{n+\frac{1}{2}}\right)\right](f)}, \tag{4.22}$$

where $\mathcal{DFT}\left[\cdot\right](f)$ is the usual DFT operator, and the correction term $e^{-j\pi f \Delta t}$ accounts for the half-time-step staggering inherent in the FDTD-method marching

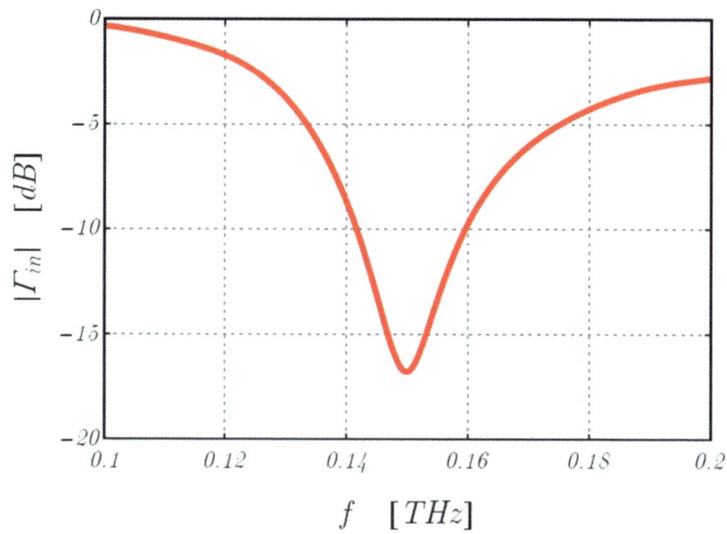

Fig. 4.5 Frequency-domain behavior of magnitude of input reflection coefficient shown by dipolelike electromagnetic sensor

algorithm. In this way, the input reflection coefficient of the sensor can be readily evaluated as

$$\Gamma_{in}(f) = \frac{Z_{in}(f) - Z_0}{Z_{in}(f) + Z_0} , \tag{4.23}$$

where $Z_0 = R_g$ denotes the reference impedance. Under the aforementioned assumptions, the magnitude of $\Gamma_{in}(f)$ features the frequency-domain behavior shown in Fig. 4.5. Using the near-field distribution as computed by the presented subcell averaging FDTD scheme in combination with the NF-FF transformation detailed in the previous section, it is straightforward to determine the far-field parameters relevant to the structure under analysis, such as the realized gain (Balanis 2005), whose angular behavior is plotted in Fig. 4.6 in excellent agreement with what is expected from theory.

In the performed numerical simulation, a ten-cell uniaxial perfectly matched layer (UPML) absorbing boundary condition (Gedney 1996) was adopted to simulate the extension of the space lattice to infinity. In particular, a quartic polynomial grading of the UPML conductivity profile was selected in such a way as to achieve a nominal value $\mathcal{R}_{PML} = 10^{-4}$ of the spurious reflection level at the truncation of the computational mesh.

4.2.4 *Integrated Small Antenna and Measurement Approach*

An integrated small RF antenna can be measured by the scheme shown in Fig. 4.7a, where a horn receiver probe reads the radiated electromagnetic field. The key

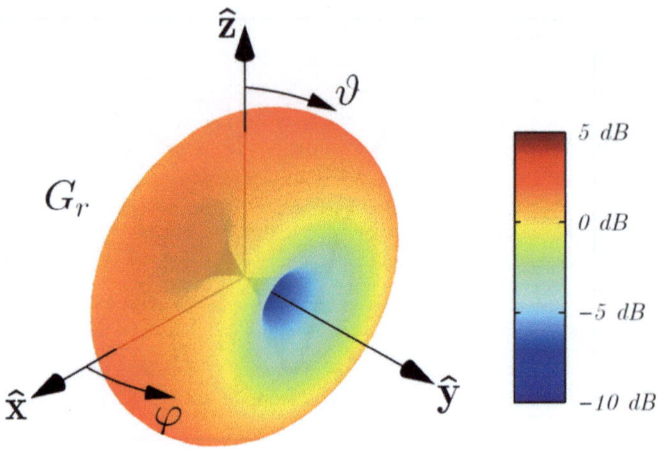

Fig. 4.6 Realized gain by dipolelike electromagnetic sensor oriented along *y*-axis of suitable Cartesian coordinate system

Fig. 4.7 (**a**) Experimental setup configuration for radiation aspects of small RF antenna. (**b**) Experimental setup configuration for measurements of S_{11} parameter

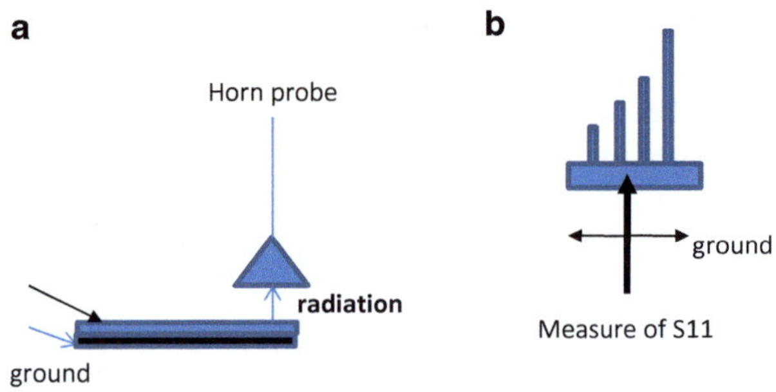

parameter S_{11}, indicating the maximum transferred power, will characterize the efficiency of the antenna. A particular layout of an antenna array is shown in Fig. 4.7b, where different waveguides (characterized by different lengths) are coupled on the same chip. An important aspect for the design of small integrated antennas is to find the thickness of the substrate layer that separates the metallic antenna or array from the metallic ground plane. The coupling of the source represents the main aspect of the design. This aspect must be considered before the fabrication of the antenna. To improve the high power transferred by the source, a coaxial cable adapter (including amplifier stage) or an integrated tapered *inplane* waveguide could be considered. The metallic layer (constituting the antenna layout) on the substrate surface can be fabricated by a RF sputtering deposition technique or a photolithography technique typically used for MEMS technology. During the fabrication process an important geometrical parameter to control is the thickness of the metallic layer. Small dimensions on the micro scale (and for nanometric metallic thickness) allow for the use of the antenna in the plasmonic range, which is characterized by a different experimental setup. The scheme of the experimental

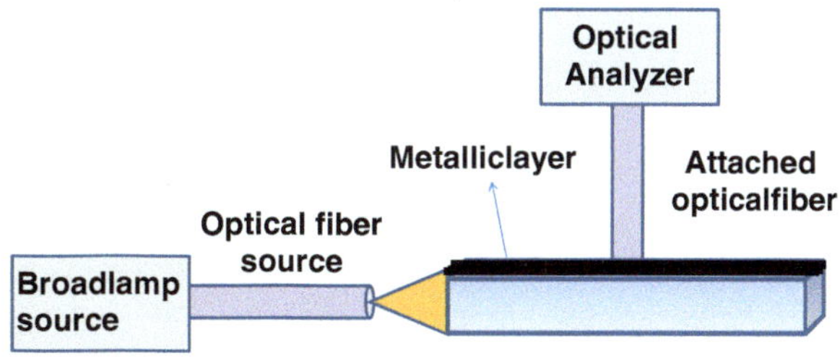

Fig. 4.8 Experimental setup configuration for radiation aspects of plasmonic antenna

setup for a plasmonic antenna configuration is illustrated in Fig. 4.8, where a lamp source coupled to an optical fiber excites *inplane* the plasmonic antenna, and another optical fiber collects the radiated plasmonic signal into an optical analyzer or into a photodetector. A plasmonic probe can be implemented as schematized in Fig. 4.9a, where a thin metallic probe is integrated on an optical waveguide. This waveguide is excited in the core region and allows a plasmonic wave to propagate as a surface wave propagating along the metallic layer (top layer). These probes are useful for sensing approach as for optical antennas. An example of an optical near-field detection system is shown in Fig. 4.9b, where a transmitter T_x metallic wedge, excited by an optical wave, radiates the focused electromagnetic field, and a receiver R_x wedge probe detects the variation in the refractive index (dielectric contrast *dn*) against a background with refractive index *n*. The distance and the orientation of both probes will be defined to distinguish the refractive index variations. Because the near field is difficult to evaluate at optical frequencies, an accurate approach that can model and design a wedge detection system is necessary.

4.3 Some Applications of Planar RF MEMS

Research in the field of radio frequency (RF) MEMS resonators has generated various techniques using periodical patterns, especially for realizing piezoelectric aluminum nitride (AlN) bandpass filters and bandgap filters. The technology and the approaches involve new classes of single chips behaving as micromechanical bandpass filters, bulk acoustic wave devices, microwaves, and millimeter-wave (mm-wave) bandstop filters. The filtering behavior is obtained using periodic substrate electromagnetic bandgap (EBG), microstrip bandgap structures, and surface modes propagating in multiple metallic layers. In particular, metallic periodic configurations are studied for the analysis of bandstop and bandpass frequency regions. The bandgaps and the cutoff frequencies can be generated by changing the electromagnetic properties of the period waveguides, for example, by applying forces on the piezoelectric material: in this case, the stress properties of the piezoelectric maZterial generate a variation in the electromagnetic field, which

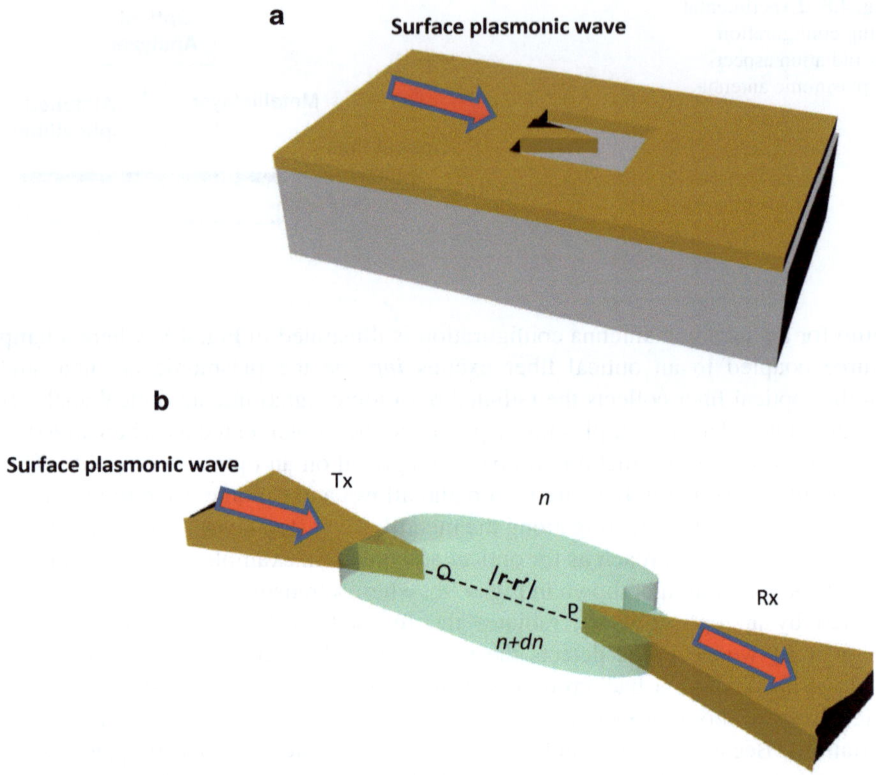

Fig. 4.9 (**a**) Layout of plasmonic probe integrated on optical waveguide. (**b**) Scheme of near-field probe detection system at optical frequencies. The increase in *dn* can be correlated to a decrease in the near-field amplitude

characterizes the guided modes of the periodic structure and the parameters of the equivalent transmission line circuit. The guided modes are defined by the working frequency and by the geometry. Several modes depending on the layout configuration can propagate along the periodic waveguides by defining the frequency response and the resonance conditions of the sensor.

4.4 Minimally Invasive Skin Cancer Wireless Detection System

An important recently investigated topic for the detection and diagnosis of melanoma is electromagnetic reflectometry. In particular, predictions of potential malignancy can be made by measuring the electrical characteristics of the skin. In vivo experiments have proved that normal skin and malignant lesions differ significantly

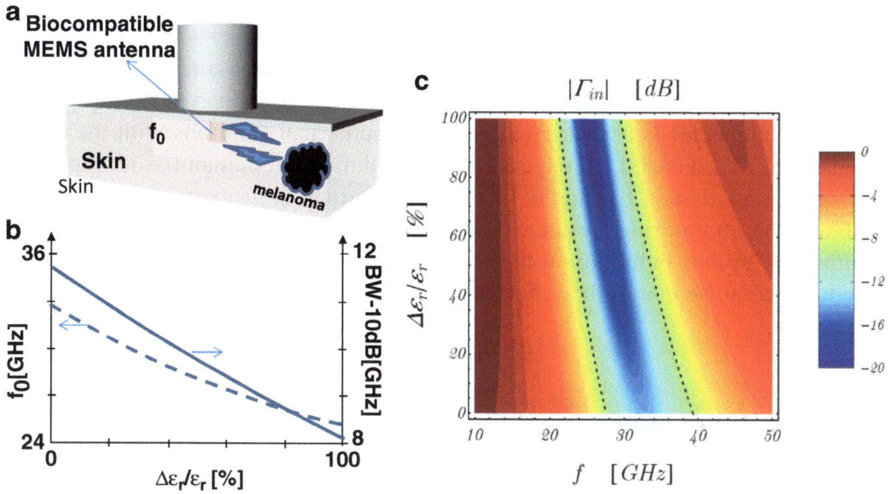

Fig. 4.10 (**a**) Operative scenario of skin melanoma detection system by 0.4-mm-long probe sensor. (**b**) Resonant frequency and bandwidth trends versus permittivity variation of skin. (**c**) Parametric risk map

in their biological properties, changing the local complex permittivity value of the tissue. Thus, an alternative way to monitor the physiological parameters characterizing tumors may be to evaluate the reflectivity response of a suitable antenna sensor. In the considered context, it is necessary to come up with an accurate design of noninvasive or minimally invasive biocompatibility small antennas that can detect anomalies in the human body. With respect to minimally invasive antennas, different biological tissues add considerable complexity to the problem due to high losses. An open-ended coaxial probe can be used to perform, in combination with a calibrated vector network analyzer (VNA), reflectometry-based measurements of various skin lesions in the gigahertz frequency band.

However, a large back-reflection level at RF power could provide an inaccurate diagnosis due to a reduced electromagnetic field interaction with the tissues being tested. To overcome the aforementioned power limitation, a piezoelectric microneedle MEMS antenna sensor for the minimally invasive detection of skin cancer could be used. This system is characterized by a good return loss and efficiency in the millimeter-wave frequency range. Furthermore, the adoption of a reconfigurable piezoelectric probe would allow for the electronic control of the near-field radiation pattern, which could focus the electromagnetic energy on the cancerous material; it would also make it possible to perform a quick spatial scan over a larger area (additional degree of freedom in the monitoring procedure). Figure 4.10a shows the scheme of a small applied probe (0.4 mm long). Figure 4.10b shows the trend of resonant frequency and bandwidth versus the permittivity variation of the skin, indicating a possible risk of melanoma, and Fig. 4.10c shows a complete map reporting all the key parameters of the risk such as the resonant frequency and the reflection coefficient. We observe that the theoretical trends are functions of the probe dimensions. Based on the piezoelectric frequency resonance, it will be possible to control the probe irradiation direction monitoring the adjacent area. A complete knowledge of the losses of human tissue at different working frequencies is necessary to correlate numerical solutions with the experimental case. To this end, an experimental simulations of human tissue (reproduction of human tissues by liquid material or animal tissues) it is significant to calibrate the monitoring system. The same approach can be employed using noninvasive probes radiating on the skin surface, properly tuning the input power to reduce the noise due to the reflectivity of the skin/air interface.

4.5 Small Antennas: Modeling Aspects

Since the near-field distribution is difficult to evaluate experimentally at optical frequencies, an accurate approach capable of modeling and designing a wedge metallic optical antenna is highly desirable. The electromagnetic field excited in the presence of metal/dielectric discontinuities features a very complex behavior that makes the adoption of detection principles impossible. To overcome this limitation, the simultaneous transverse resonance diffraction (STRD) method can be conveniently used to evaluate the near-field distribution excited by an infinitely thin metal conductor (in the case of a plasmonic layer). In this way, field singularity issues arising in radiation problems involving optical antennas can be efficiently handled. The traditional numerical approaches such as the FEM and FDTD technique are computationally expensive when calculating the radiated field in the proximity of a dielectric/metallic wedge. Accurate methods based on the multipole expansion of Green's function (MEG) combined with the novel principle of STRD can be used for the evaluation of a radiation pattern. In particular, the STRD/MEG method is suitable for the plasmonic resonance of nanoprobe systems and wedge conductors. Theoretically speaking, a small antenna whose length L is $\lambda_0/5 < L < \lambda_0/10$ (λ_0 is the working wavelength) behaves as an infinitesimal dipole (small dipole). Figure 4.11 illustrates an example of an infinitely thin metal probe. The radiated electric field of a small antenna at an external point $\mathrm{P} = \mathrm{P}(\mathrm{r},\theta,\varphi)$ can be evaluated by the source wedge point at $\mathrm{Q} = \mathrm{Q}(\mathrm{r}',\theta',\varphi')$. In spherical coordinates, the Green's function of satisfies the Helmholtz scalar equation and the multipole expansion of the Green's function given by

$$G(\mathbf{r},\mathbf{r}') = \sum\nolimits_{l,m} f_l(r,r')T^{*}_{l,m}(\theta',\varphi')T_{l,m}(\theta,\varphi), \tag{4.24}$$

where the functions $T_{l,m}$ are defined by Legendre functions, and the terms f_l are defined by spherical Bessel and spherical Hankel functions.

We observe that (4.1) presents a discontinuity peak of the first derivative at the point Q (point of singularity). According to this property, the MEG is suitable for

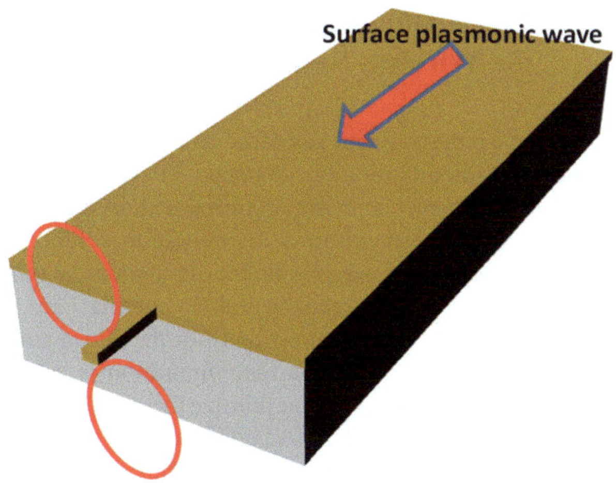

Fig. 4.11 Infinitely thin metal conductor excited by surface plasmon wave

near-field evaluation in the proximity of metallic singularities. Moreover, the index l indicates the order of the expansion and represents the accuracy of the solution: the convergence of the near-field solution can be performed by increasing the order of MEG. The solution of Maxwell's equations in cylindrical coordinates is provided by the following three field components:

$$\begin{aligned} V_\rho &= -j\omega\mu\frac{1}{\rho}\frac{\partial\Psi}{\partial\phi}, \\ V_\phi &= j\omega\mu\frac{\partial\Psi}{\partial\rho}, \\ I_z &= k^2\Psi, \end{aligned} \tag{4.25}$$

where Φ is a suitable Hertzian potential that satisfies the Helmholtz equation. The field in (4.24) is transverse electric (TE) with respect to the z-direction, denoted by TE^z, but it is also transverse magnetic (TM) with respect to the ϕ-direction (thus denoted TM^ϕ). A similar modal analysis can be carried out for TM^z (i.e., TE^ϕ) fields characterized by the following components:

$$\begin{aligned} I_\rho &= j\omega\varepsilon\frac{1}{\rho}\frac{\partial\Psi}{\partial\phi}, \\ I_\phi &= -j\omega\varepsilon\frac{\partial\Psi}{\partial\rho}, \\ V_z &= k^2\Psi. \end{aligned} \tag{4.26}$$

Equations (4.25) and (4.26) are used to calculate the total electric field near the metallic wedge using the relationships that convert the cylindrical coordinates into spherical ones.

4.6 Three-Dimensional Micrometric Antenna for Wireless Systems

Currently, industrial and scientific research communities are moving on the development of reconfigurable antennas. In particular, the interest is in the various functionalities such as radar, communication, direction and spectrum, and control. Moreover, personal wireless or vehicle-to-vehicle communication devices must typically support a large number of standards (e.g., UMTS, Bluetooth, WiFi, WiMAX, DSRC). A reconfigurable antenna can reuse its volume at different frequency bands so that a portion of or the entire structure is involved in a specific mode of operation. Antenna reconfiguration is achieved by changing the state of suitable switching devices to optimize the performance of the device for applications involving control signals. Most approaches to reconfiguring antennas are developed by electromechanical switches and include switches based on PIN diodes, varactors or field-effect transistors (FETs), and small piezoelectric antennas. A 3D view of a small antenna geometry is shown in Fig. 4.12a, b. In the proposed design, the radiating spiral is described by the different types of microelectromechanical system (MEMS). In the presented section, a suitable solid-state tuning circuitry is used to dynamically adjust the circuital characteristics, such as the frequency of operation (the antenna works in the terahertz range), or the radiation properties of a novel nonconformal spiral multiport 3D microantenna for next-generation wireless communications. The device reconfiguration is technically achieved by changing the feeding/loading condition at the input ports and, consequently, the current distribution within the antenna volume. The topology of the proposed nonconformal Fermat-like spiral microantenna is characterized by an n-turn spiral arm with a width of $w_l = 500$ nm. The antenna is printed on a dielectric hemisphere with a diameter of $D_s = 40$ μm.

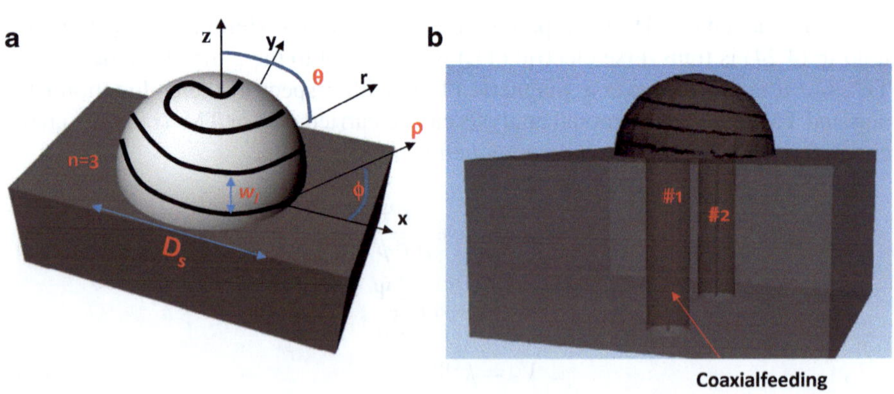

Fig. 4.12 (**a**) 3D view of reconfigurable spherical Fermat spiral multiport antenna. The substrate-integrated coaxial feeding lines used to excite the radiating structure are shown. (**b**) 3D CAD design

Acknowledgments The author gratefully acknowledges the help of Diego Caratelli.

References

Balanis, C.A.: Antenna Theory: Analysis and Design, 3rd edn. Wiley-Interscience, Hoboken (2005)

Caratelli, D., Cicchetti, R.: A full-wave analysis of interdigital capacitors for planar integrated circuits. IEEE Trans. Magn. **39**(3), 1598–1601 (2003)

Caratelli, D., Yarovoy, A.: Unified time- and frequency-domain approach for accurate modeling of electromagnetic radiation processes in ultra-wideband antennas. IEEE Trans. Antennas Propagat. **58**(10), 3239–3255 (2010)

Chen, F.C., Chew, W.C.: Time domain ultra-wideband microwave imaging radar system. J. Electromagn. Waves Appl. **17**(2), 313–331 (2003)

Felsen, L.B., Marcuvitz, N.: Radiation and Scattering of Waves. Prentice-Hall/IEEE Press, New York (1994)

Gedney, S.D.: An anisotropic perfectly matched layer-absorbing medium for the truncation of FDTD lattices. IEEE Trans. Antennas Propagat. **44**(12), 1630–1639 (1996)

Jin, J.M.: The Finite Element Method in Electromagnetics, 2nd edn. Wiley, New York (2002)

Massaro, A., De Guido, S., Ingrosso, I., Cingolani, R., De Vittorio, M., Cori, M., Bertacchini, A., Larcher, L., Passaseo, A.: Free standing piezoelectric rings for high efficiency energy harvesting at low frequency. Appl. Phys. Lett. **98**, 053502 (2011)

Mukhopadhyay, S.C.: Novel planar electromagnetic sensors: modeling and performance evaluation. Sensors **5**, 546–579 (2005)

Ramahi, O.M.: Near- and far-field calculations in FDTD simulations using Kirchhoff surface integral representation. IEEE Trans. Antennas Propagat. **45**(5), 753–759 (1997)

Stratton, J.A.: Electromagnetic Theory. IEEE Press, Piscataway (2007)

Stutzman, W.L., Thiele, G.A.: Antenna Theory and Design, 2nd edn. Wiley, New York (1997)

Taflove, A., Hagness, S.C.: Computational Electrodynamics: The Finite Difference Time Domain Method, 3rd edn. Artech House, Norwood (2005)

Yee, K.S.: Numerical solution of initial boundary value problems involving Maxwell's equations. IEEE Trans. Antennas Propagat. **14**(3), 302–307 (1966)

Chapter 5
Materials and Innovative Systems Oriented to the Implementation of Research Platforms Involving Devices and Electronic Systems

Abstract In this chapter, we define new classes of materials such as nanocomposite materials (NMs) and plasmonic materials. Applications in nanoelectronics, micro-electromechanical systems (MEMS), nanoelectromechanical systems (NEMS), energy harvesting from the human body, and for implanted systems are presented. We discuss the possibility of using innovative materials and systems to create a research platform. Some suggestions about the facilities necessary to create a research platform are noted. The goal of this chapter is to show how to merge research in electronics and optics to spur some thinking on how to formulate a good project based on applications.

5.1 Introduction to Nanocomposite Materials

Polymers such as polydimethylsiloxane (PDMS), Su-8, polystyrene (PS), poly(methyl methacrylate) (PMMA), polyvinylidene fluoride (PVDF), polyethylmethacrylate-comethylacrylate (PEMMA), TOPAS, polyvinyl alcohol (PVA), polyaniline, chitosan, and polycarbonate (PC) are currently and commonly used in many applications including tactile sensing, minimally invasive microsurgeries, the military, and industrial automation systems. Metallic/organic nanostructures added to polymeric materials generate nanocomposite materials (NMs). NM applications can be implemented in RF/microwave, radiation applications, biosensing, and piezoelectric applications. Because some NMs are conductive, they can be used as micro-/nano probes for detection systems. To understand the type of implementation, a characterization of the NMs is necessary. In fact, nanofillers could improve an high sensitivity orienting the applications (e.g., temperature sensing, pressure sensing, gas sensing). If nano objects are generated on a substrate surface (electrospun nanofibers, autoassembled nanofibers, textured nanopatterns), then the study of the surface properties can be useful for wettability and microfluidic problems. Before characterizing the chemical/physical properties of NMs it is essential to assure the repeatability of the fabrication process.

A. Lay-Ekuakille, *Optical Waveguiding and Applied Photonics*, Lecture Notes in Nanoscale Science and Technology 10, DOI 10.1007/978-1-4614-5959-0_5,

Micro-/nanofillers can drastically change the electrical, optical, piezoelectric, and chemical properties of the initial polymeric matrix. Thus, the tuning of the material properties can be performed by controlling the filler generation. With respect to the optical properties, gold (Au) nanoparticles generated by a chemical reduction in PDMS (by means of a $HAuCl_4$ gold precursor) facilitate light coupling between a tapered optical fiber and a PDMS-Au NM deposited on the core surface (Massaro et al. 2011). The pressure applied to the PDMS-Au NM will change the effective permittivity: a major density of gold nanoparticles will increase the effective permittivity, and consequently, the light will be lost to the PDMS-Au NM, allowing for a decrease in the light intensity transmitted at the output of the fibers. Electrospun fibers can increase the interaction area for gas sensing or trapping (in this case, the sample should be placed in a closed chamber to estimate the variation of the electrical conductivity due to the gas interaction). A basic parameter of NMs is the capability of the material to be reused. PDMS-Au NMs have the ability to return to their initial mechanical and optical configurations (without applying pressure forces), and polymers reinforced by nanofillers do not create internal cracks during the testing process. A sensor based on gold micro-/nanoparticles embedded in a polymer film may enhance the radiated optical field due to light scattering and coupling, providing a good sensor efficiency. Optical properties can be tuned by changing the concentration of metallic/organic/inorganic oxides micro/nano inclusions. In particular, an analytical study of the control of the filler distribution and its effect on the resulting properties and performance of nanocomposites could help researchers during the design process. The polymer–filler interaction, which seems to be a key point for filler distribution, is an important issue to analyze. Fillers are expected to interact strongly with polar functional groups in polymers. An increased understanding of the involved mechanisms would make it possible to achieve the desired performance of the polymer composites, such as reduced actuation voltage, increased electroactive strain, and improved response time. Other solutions involving piezoelectric NMs entail the introduction of nanofibers or inclusions in different polymeric materials such as PVDF and PDMS. The dual goal of piezoelectric NMs is to characterize simultaneously mechanical and piezoelectric properties. The studies will be planned by designing a packaging compatible with the whole device/system. An innovative application involving optical fiber sensors following the physical principle described in Massaro et al. (2011) could be to have an optical fiber embedded in a NM that would detect variations in the water pressure due to a broken pipeline or to changes in the water flow. This sensor could detect variations in the transmitted intensity due to applied pressure forces. In this case, the pressure force is generated by a crack in the pipeline or by variations in the water flow. Due to the low cost of optical fibers, optical fiber sensors can be applied in different parts of a pipeline network for complete monitoring. Optical fiber sensors can be implemented in wireless systems to create an efficient and real-time detection system. Another application involving NMs might consider an optical waveguide sensor capable of detecting the pollution degree of water. This second class of sensor is based on the concept of guided optical modes of a NMS waveguide and can be combined with the previous optical fiber sensor measuring the variation of the water permittivity due to the different kinds of pollution (e.g., oil, natural pollution): different kinds of pollution or concentration will change the permittivity of the waveguide cladding, which in turn will change the mode properties at the output.

5.1.1 Approach to Studying Nanocomposite Materials for Solar Cells

Today NMs are also used in solar cells. Solar cells represent a typical example of optoelectronic devices. The study of this topic must take into account both the optical and optical aspects.

Nanocomposite systems are typically on the order of nanoparticles some tens of nanometers or more in size dispersed in a polymeric matrix and require the use of modeling approaches based on classical electromagnetism or analytical optical methods to describe the electromagnetic response. A quantum mechanical description, however, may provide, as in the Effective Mass Approximation (EMA), a detailed description of the electronic levels involved in optical transitions. Keeping the solar spectrum as a reference for absorption frequencies, it is possible to optimize at the quantum level the absorbing gap by varying the geometrical characteristics of the system, such as the (metallic/organic) nanoparticle radius, interparticle distance, and shape distribution. Optimized systems must be passed to tools for absorption and luminescence calculations within classical limits. Given the dimensions of investigated systems, parallelized tools could optimize the efficiency and could lead to high-performance architectures.

5.1.2 Electrical Conductive Micro-/Nanoparticles in Nanocomposite Materials

Conductive micro-/nanoparticles are an important issue in the implementation of small sensors based on electrical conduction (Kumar et al. 1998). The choice between bulk material loss and the surface boundary condition is problem dependent. The boundary condition should be applied when the conductor is much thicker than skin depth at the solution frequency. If the conductor is not thick relative to skin depth, the bulk material conductivity must be used to arrive at an accurate solution. The wave equation that characterizes the conductivity of a micro-/nanoparticle will be defined as follows:

$$\nabla \times \left(\frac{-1}{j\omega\mu} \nabla \times \mathbf{V} \right) = (j\omega\varepsilon + \sigma)\mathbf{V}. \tag{5.1}$$

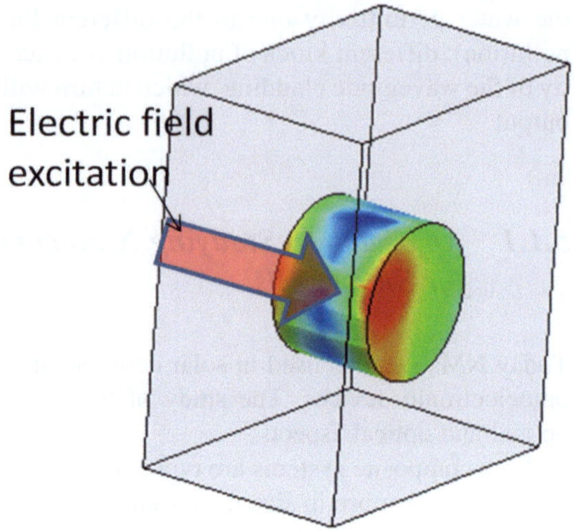

Fig. 5.1 Schematic illustration of volumetric distribution of density of current $\mathbf{J}_{vol}$ of a gold cylindrical microparticle excited by an external electric field and embedded in a box bulk cell

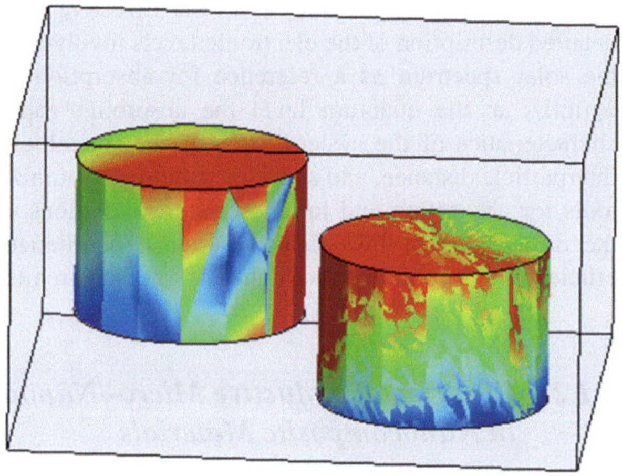

Fig. 5.2 Schematic illustration of calculated volumetric distribution of density of current $\mathbf{J}_{vol}$ of electrically coupled gold cylindrical microparticles excited by an external electric field

Figure 5.1 shows the schematic density of the current $\mathbf{J}_{vol}$ calculated by considering a gold cylindrical rod embedded in air (the air box that embeds the metallic particle is excited by an external electric field exciting a port). More metallic cylindrical pillars, embedded in a polymeric material, generates a nanocomposite conductive material, changing the electrical property of the pure polymeric matrix. The effect of the electrical coupling can be studied by analyzing the density of the current coupling for different pillars placed nearby. In Fig. 5.2 we show an example of coupled gold pillars. Experimentally it is difficult to measure the conductivity of a single micro-/nanorod. For this reason, it is preferable to measure the conductivity of a bulk polymer matrix embedding different rods. The bulk characterization will be defined by the measured conductivity of the NM for different concentrations of small micro-/nanofillers. Numerically, it is better to

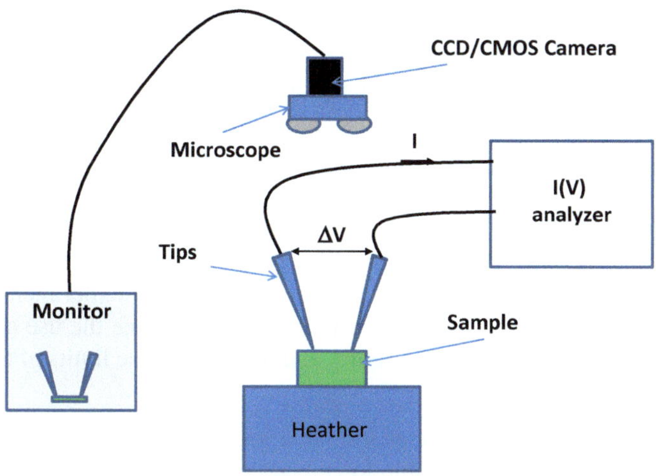

Fig. 5.3 Scheme of an experimental setup for the characterization of a temperature sensor

simulate a unit bulk cell by changing the number of rod fillers (the number of rod fillers will define the filling factor). The unit bulk cell is constituted by a box of a polymer that can enclose metallic particles. The behavior of a single cell will be scaled for a large bulk dimension to simulate a millimeter-scale nanonocomposite sample. The simulation of conductive nanocomposite cells will be useful for the design of innovative solar cells or optoelectronic sensors based on electrical conductivity. With respect to the solar cells, another parameter to consider is the solar light absorption. Light absorbance is defined as $A = -\log 10(I_T/I_0)$, where I_0 is the input intensity and I_T the transmitted intensity. The absorbance represents the resonant energy of the nanocomposite bulk and the electromagnetic energy (represented by the electrical field **E**), which can be enhanced. This simultaneous effect (light absorption and electrical conductivity) is correlated by the constitutive relationship $\mathbf{J} = \sigma\mathbf{E}$. The electrical conductivity of a nanocomposite sample can be measured by a probe station composed of a digital current/voltage I(V) reader, by sensitive controlled tips (able to transfer a ΔV signal), and by a camera connected to a microscope and to a monitor capable of checking the exact position of the tips on the sample to measure. An important application of NMs is the effect of temperature on electrical conductivity. Highly sensitive material versus the temperature could be used as temperature sensors in environments where it is important to detect small variations in temperature for the equilibrium of a system. The sensitivity can be tuned by acting on the metallic/organic filler concentration. A typical experimental setup for temperature characterization is shown in Fig. 5.3. In this scheme a controlled heather changes the electrical conductivity of the bulk sample, and two tungsten tips measure the I(V) characteristics (setting the diode modality of the digital analyzer) as a function of temperature.

Conductivity can be studied by applying voltage signals on electrodes that enclose the sample: the current–voltage characteristic I(V) will be estimated by means of

Ohm's law, the resistivity, and, consequently, the conductivity of the sample. The probe station typically used for electrical measurements is shown in Fig. 5.16a.

A good protocol for the temperature characterization of a nanocomposite sample is described as follows:

- Preparation of different nanocomposite samples having the same dimensions and different concentration of metallic/organic fillers (tuning of the filling factor)
- Measurement of I(V) characteristics for different values of the temperature and maintaining the same distance of the tips on the bulk surface
- Taking average of different measurements (reduction of experimental error)
- Analysis of sensitive temperature range in order to define the use of the sensor (high temperature or low temperature) and to establish the limits of the analyzed material

To measure the I(V) characteristic of a nanocomposite sample without an applied temperature, it is preferable to embed the nanocomposite sample between two metallic plates defining a capacitance: this layout allows one to measure low currents where nonconductive polymers are considered. Fillers change the electrical properties of polymers by increasing the conductivity. This electrical property can be modified by tuning the concentration of the inclusions and successively measuring the I(V) characteristic (the currents of NMs are a function of the filling factors (concentration of the inclusions)). The increase in the conductivity is an important specification, especially for gas sensors, where a high sensitivity versus conductivity is required.

5.2 Surface Plasmons and Materials

The excitation of surface plasmons (or plasmon polaritons) by light is called the surface plasmon resonance (SPR). The SPR occurs for nanometer-sized metallic planar surfaces (Simon et al. 1975) or is localized. Typically, gold and silver surfaces are used for the characterization of SPR. Gold materials are particularly easy to use in SPR biosensor applications. Surface plasmons are surface electromagnetic waves propagating in a direction parallel to the metal/dielectric (or metal/vacuum) interface. Various models, such as, for example, quantum theory or the Drude model, are actually used to describe the existence and properties of surface plasmons. The simplest way to approach the problem is to consider each material as a homogeneous continuum described by a frequency-dependent relative complex permittivity: the real part of the dielectric constant of the metal must be negative and its magnitude must be greater than that of the dielectric constant. This condition is encountered in the IR-visible wavelength region for air/metal and water/metal interfaces (where the real dielectric constant of a metal is negative and that of air or water is positive). To excite resonant surface plasmons, an electron or light beam can be used. The incoming beam should match its impulse to that of the plasmon. P-polarized light (polarization parallel to the plane of incidence) achieves the resonance at a given wavelength and angle. In addition, S-polarized (polarization perpendicular to the plane of incidence) light cannot excite surface plasmons. Very thin metallic layers could be used as uniform layers allowing the propagation of surface plasmons. The high density of metallic nanoparticles constituting a monolayer can be generated on a substrate surface by means of advanced lithographic techniques or by advanced chemical processes using gold or silver precursors and polymeric material as substrates (such as chitosan or PDMS). The type, dimensions, and shape of the generated nanoparticles will determine the peak of the absorbance corresponding to the plasmonic resonance. For example, quasi-spherical gold nanoparticles are characterized by a plasmonic resonance of around $\lambda = 530$ nm. Increasing the density (provided by the concentration) of the nanoparticles will lead to an increase in the absorbance peak intensity. Agglomerates or shapes such as crystals or rods may change the plasmonic wavelength resonance. Another technological approach used to fabricate thin metallic plasmonic films is the RF sputtering deposition technique, which makes it possible to design controlled thickness of the metallic layers (thickness is a function of the input power, the deposition time, gas pressure, and temperature, and the adhesion of the metal is also a function of the type of substrate).

5.3 Nanoelectronics for Implanted Systems

Electronic nanocomponents for sensors, wireless systems, MEMS, and power management integrated circuits (ICs) could enable the commercial rollout of the next generation of innovative electronic devices and systems. Applications include home automation, building automation, industrial process/automated meter reading, medicine, the military, automotive tire pressure sensors, radio frequency identification, and others. In particular, biomedical implants provide diagnostic information to enable personalized medicine. Batteries meet all of the energy needs of implants, but batteries inherently limit the operating lifetime of implants due to their fixed energy density and the strict volume and weight constraints imposed by implantable devices. For this reason miniaturized electronic components could increase the efficiency of the implanted devices. Moreover, the use of micro-/nanoelectronic components might miniaturize chips for wearable sensing in physiological activities of the human body, such as, for example, blood pressure, temperature, glucose status, and brain activity. Another important issue in nanoelectronic components concerns the fabrication of innovative tuning circuits for small antennas able to detect human tissue anomalies and innovative miniaturized labs on chips. A multidisciplinary study involving the combined efforts of mechanical engineers, electrical engineers, and clinicians is essential for synthesizing reliable, integrated, and biocompatible circuits that might extend the lifetime and functionality of current implants and enable a new generation of biomedical devices. A worthwhile project would consider the compatibility of such components with currently available devices and would be able to supply energy to different apparatus and instruments for daily use.

5.3.1 Nanoelectronics and Nanocomposite Materials

In the past decade, nanocomposite or hybrid materials have gained increasing attention for nanoelectronics applications. Indeed, these materials exhibit unique properties of interest in many fields of application because they combine several properties due to their intrinsic compositions. By combining specific polymers and fillers, NMs can be highly interesting due to their electrical conductivity for different orders of currents. In particular, the use of NMs can be oriented toward future sensing applications where conductivity enhancement will be required. One strategy is based on the use of gold nanoparticles introduced in transparent polymer films and on the generation of tailored areas with a controlled nanopatterning (e.g., aligned gold nanoparticles, localized graphene nanoplatelets, or carbon nanotubes). Because the electrical properties of NMs depend on the gold nanoparticle density and size, a main aspect for the design on nanoelectronic components is the control of the nanoparticle generation process. Natural polymers can be used as biocompatible substrates for conductive gold nanopatters.

5.3.2 Expected Impact of Nanoelectronic Implanted Devices and Basic Procedures to Create a Research Platform

Nanoelectronic implanted systems contribute in application areas such as (i) public medical infrastructures; (ii) medicine (surgical equipment, diagnostic equipment, imaging equipment, health monitoring devices, systems and equipment). Furthermore, nanoelectronic components make it possible to harvest energy from human bodies, (ii) reduce time to market despite the increasing contribution of embedded systems and software and their increasing size and complexity, and (iii) increase the quality and reliability of products and services while providing novel functionalities to the user. A worthwhile project would include tools as software, RF-microwave detectors, new detection techniques, packaging, and technology, which can be considered supporting products in the biomedical industry. The design of modular tools could allow the use also in veterinary medicine.

A modern approach can reduce costs for industry and business, create markets for environmental goods and services, drive innovation, reduce business risk and increase the confidence of the investment markets and insurers, create competitive advantage and competitive markets, create and sustain jobs, improve the health of the workforce and the general public, and protect the natural resources that businesses and we all depend on. The properties and functions of novel new structures, devices, and detection systems are directly tied to their atomic, molecular, or macromolecular size. As progress in the manufacture and characterization of micro-/nanoscale materials continues to accelerate, a growing list of stakeholder needs has arisen. Tests are needed to support micro-/nanoscale measurement and characterization and to show how the new technology will impact public health and safety.

The main goals of a project for a startup research platform are as follows:

- Coordinate and provide a forum for academia, individual industries, standards developing organizations, and governmental entities to define needs, determine work plans, and establish priorities for updating standards or creating new standards
- Solicit participation from nanotechnology-related sectors and academia that have not traditionally participated in voluntary standards systems, and work cooperatively to achieve the mission of the project
- Facilitate the timely development and adoption of standards that are responsive to identified needs in the area of nanotechnology in general and nomenclature/terminology in particular
- Facilitate and promote cross-sector collaborative efforts between standards developing organizations to establish work plans and develop joint or complementary standards
- Where standards do not exist, obtain agreement from a standards developer to initiate the development of standards in a timely manner
- Establish and maintain relations with other national, regional, and international standards efforts to address micro-/nanotechnology issues so as to create identical or harmonize existing standards
- Establish and maintain a database of micro-/nanotechnology electronic standards that is available on the Internet and capable of generating updates, notices, and reports
- Identify any obstacles preventing the timely adoption of needed national standards
- Make the results of projects widely available
- Provide full access to common databases
- Define flexible protocols for the standardization of new nanoelectronic devices, approaches, and medical techniques
- Adaptation of market (facilitate the introduction of innovative nanoelectronic devices for the human body)

5.4 Harvesting Energy from the Human Body: Technology Improvements and Related Sensing Issues

5.4.1 Introduction

Apart from the energy produced by means of vibrations and movements, human body heat is also a matter of interest because it is an almost and a permanent source of energy to be exploited. The human body is subject to the same laws of physics as other objects, gaining and losing heat by conduction, convection, and radiation. Conduction happens between bodies or substances in contact; convection involves the transfer of heat from a warm body to a body of air above it or inside a human body, where blood, gases, or other fluids constitute the medium; and radiant heat transfer is a major mechanism of thermal exchange between the human body and the surrounding surface environment. In most situations, these three effects—conduction, convection, and radiation—operate together. In the human body, metabolic processes generate their own heat as well, much like a heat-producing engine. The human body strives to be in a stable state, so it absorbs and emits energy to attain equilibrium; stimulation is applied to the body surface; this make the activity of metabolism induced to body surface.

Human beings and, more generally, warm-blooded animals (e.g., some dangerous and endangered animals, cattle, and pets) can also be heat sources, by means of a thermoelectric generator (TEG), for devices attached to their skin. A TEG mounted in a wristwatch is an example of how a watch can be powered by wasted human heat. In various works, changes in a part of the human body were studied and analyzed before and after stimulation and comparisons were made between them; then the bioheat transfer mechanism was simulated using a second-order circuit designed on the basis of a first-order circuit introduced by Guotai et al. (2004), and the human thermo response was analyzed. Jiang et al. (2001) studied the diagnosis of diabetes by thermograph. Since the human body emits energy as heat, it is desirable to try to harness this energy. However, Carnot efficiency puts an upper limit on how well this waste heat can be recovered. Assuming a normal body temperature and a relatively low room temperature (20 °C), the Carnot efficiency is (Ekuakille et al. 2009)

$$\frac{T_{body} - T_{environment}}{T_{body}} = C1 = 5.5\%. \tag{5.2}$$

In a hot environment (27 °C) the Carnot efficiency falls to

$$\frac{T_{body} - T_{environment}}{T_{body}} = C2 = 3.2\%. \tag{5.3}$$

This calculation provides an ideal value. Today's thermoelectric generators that might harness this energy do not approach Carnot efficiency in energy conversion. Although work on new materials and new approaches to thermoelectric generators promise to somewhat improve conversion efficiencies, today's standard thermopiles are 0.2–0.8 % efficient for temperature differences of 5–20 °C, as expected for a wearable system in temperate environments. For the sake of discussion, the theoretical Carnot limit will be used in the analysis below, hence the numbers are optimistic. While sitting, a total of about 116 W of power is available. Using a Carnot engine to model the recoverable energy yields 3.7–6.4 W of power. Under more extreme temperature differences, higher efficiencies may be achieved, though robbing users of heat in adverse environmental temperatures is not practical. Evaporative heat loss from humans accounts for 25 % of their total heat dissipation (basal, nonsweating) even under the best of conditions. This "insensible perspiration" consists of water diffusing through the skin, with the sweat glands keeping the skin of the palms and soles pliable, and the expulsion of water-saturated air from the lungs. Thus, the maximum power available without trying to reclaim heat expended by the latent heat of vaporization drops to 2.8–4.8 W.

5.4.2 *Objectives*

A good project has different subobjectives that lead to the most important one: extracting any form of energy from the human body to supply autonomous devices for diverse purposes and applications as biomedical devices, apparatus and instruments for daily use, and others. Vibrational energy could be extracted from breathing, walking, and eating while heat energy is extracted from skin.

5.4.3 *Piezoelectric Issues*

Many types of piezoelectric materials display piezoelectric properties. Some are naturally occurring materials, such as quartz, while others have been engineered. We are investigating using lead zirconate titanate (PZT), which exhibits a relatively high conversion of mechanical to electrical energy. Current piezoelectric energy-harvesting research falls into two key areas: developing optimal energy-harvesting structures and highly efficient electrical circuits to store the generated charge or present it to the load circuit. Our research focuses primarily on the first area, in which the goal is to create small, lightweight structures that couple very well to mechanical excitation and convert the most useable electrical energy.

Embedded deployments, such as sensor-actuator networks, constitute a large class of pervasive computing devices. Unlike cell phones or laptops, which users can periodically recharge, pervasive devices must operate on their initial batteries. The highest reported energy densities for current battery technologies range around 3.78 kJ/cm3, which implies that for a low-power device operating at an average consumption of 1 mW to have a 10-year lifespan, it needs a large 100 cm^3 battery. Thus, energy supply is a major bottleneck for system lifetime, and harvesting energy from the deployment environs can help alleviate this.

Today's commercial micropower analog operational amplifier systems enable simple analog sensor processing at under a few microwatts, and nanopower comparitors enable digital discrimination of analog signals for less than a microwatt. Exploiting these advances, we it is worth built a family of devices allowing a node to become activated directly by analog sensor stimuli conditioned by passive or ultra-low-power active filtering.

5.4.4 Nanocomposite Piezoelectric Issues

The polymer–filler interaction appears to play a key role in filler distribution. Fillers are expected to interact strongly with polar functional groups in polymers. A deep understanding of the involved mechanisms is of crucial importance in achieving the desired piezoelectric performance in polymer composites, such as reduced actuation voltage, increased electroactive strain, and improved response time. Alternative design solutions will also be investigated. Such solutions are based on the inclusion of PVDF fibers in rubber materials commonly used for human body activities, as well as on different piezoelectric inclusions in a polymeric matrix such as PDMS. The twofold goal of a complete research project would be to characterize simultaneously the mechanical and the piezoelectric properties of innovative NMs for energy-harvesting applications.

5.4.5 Power Management

A key consideration that affects power management in an energy-harvesting node is that instead of minimizing energy consumption, a better design objective may be to operate in an energy-neutral mode, consuming only as much energy as is harvested. This strategy can enable an indefinitely long lifetime, limited only by the hardware longevity. Reducing power consumption below the level needed for energy neutrality will not increase the lifetime any further.

5.4.6 Sensory Network Issue

If one consider a distributed network in which some or all the nodes have a harvesting opportunity. The performance potential from the energy environment in which the network is situated depends on the spatiotemporal variation in the energy availability across the network. The distribution of this energy in space and time significantly affects network performance. For instance, if large amounts of energy are available but concentrated only in a small region of the network, the nodes in regions without an energy supply will limit the total useful lifetime of the network, beyond which any available energy in other regions may not be useful.

5.4.7 Modeling of Implantable Antenna Sensors

Demand for the use of RF antennas inside/outside human bodies has increased for biomedical applications. Most of the research on antennas for medical applications has focused on producing hyperthermia for medical treatments and monitoring various physiological parameters. Antennas used to elevate the temperature of cancer tissues are located inside or outside a patient's body, and the shapes of antennas used depend on their locations. For instance, waveguide or low-profile antennas are externally positioned, whereas monopole or dipole antennas transformed from a coaxial cable are designed for internal use. In addition, in medical therapy and diagnosis, telecommunications are regarded as important functions for implantable medical devices, which need to transmit diagnostic information. In contrast to the number of research accomplishments related to hyperthermia, work on antennas used to build communication links between implanted devices and exterior instruments for biotelemetry is not widely reported. It is commonly recognized that modern wireless technology will play an important role in making telemedicine possible. In the not-too-distant future, remote health-care monitoring by wireless networks will be a feasible treatment for patients affected by disease. To establish effective and efficient communication links for biomedical telemetry devices, it is crucial to give special attention to the antenna sensor. The design of implantable antennas is very challenging due to the demanding requirements in terms of reduced volume occupation, good impedance matching, low power consumption, and biocompatibility with the human body. Furthermore, the host environment, consisting of different biological tissues, adds significant complexity to the problem due to high losses. In particular, the evaluation of the circuital characteristics and radiation properties of complex metal-dielectric antenna structures requires accurate and fast numerical schemes for solving Maxwell's equations. One such technique is the finite-difference time-domain (FDTD) algorithm. However, in the conventional formulation proposed by Yee, each cell in the computational grid adopted for carrying out electromagnetic field computations is implicitly supposed to be filled by a homogeneous material. For this reason, the adoption of well-posed orthogonal Cartesian meshes could be responsible for a reduced numerical accuracy when complex electromagnetic structures having nonconformal geometries must be modeled. In such cases, dedicated locally conformal FDTD procedures to be developed within this work package provide clear advantages over the use of the staircasing approach or unstructured and stretched space lattices that might suffer from significant numerical dispersion or instability. Furthermore, in this way a detailed and insightful understanding of the physical mechanisms responsible for the performance of implantable antenna sensors can be achieved. This in turn is very important in order to enhance device reliability, optimizing the design cycle of biomedical systems for telemetry applications.

5.4.8 Implantable Antenna Sensors to Be Integrated with Signal Processing Units and Thermoelectric Converters of Human Body Heat

The human body is subjected to heat transfer by conduction, convection, and radiation. Conduction takes place between the body and other objects in close contact; convection involves the transfer of heat through blood, gases, and other fluids; finally, the radiant heat transfer is the major mechanism in thermal exchange between the human body and the surrounding environment. These three effects occur at the same time. Furthermore, all metabolic processes result in the generation of heat, much like what occurs in conventional engines, which can be usefully harvested to supply implanted biomedical devices for monitoring physiological parameters such as temperature and blood pressure. That can be easily achieved using thermoelectric microconverters, thereby avoiding the use of batteries or percutaneous energy-supplying systems. In fact, implantable batteries are characterized by their large size and finite lifetime, which requires periodic surgery. Furthermore, potential leaks from batteries pose a serious health hazard, which makes such an approach nonviable. On the other hand, percutaneous transfer has the disadvantage that wires cause a permanent breach of the skin's natural barrier to bacterial infections, which are common complications of such implants. As a consequence, the adoption of thermoelectric converters of human body heat is surely to be preferred in the design of long-term implantable devices. They can be easily integrated with different processing units devoted to the monitoring of various biological signals, which are in turn transmitted to the external equipment by means of a suitable antenna structure. It is worth noting that the antenna can also provide certain circuit functions such as resonating, filtering, and duplexing, in addition to its original role as a radiating element. In other words, a biomedical sensor for telemetry applications may be regarded as an active microwave circuit in which the output port is an antenna embedded in the human body instead of the conventional 50 Ω interface. As can be easily inferred, in this context, the loading effect of the antenna on the signal-processing circuitry must be carefully investigated. Typically, electromagnetic field solvers and measurement systems, such as network analyzers, generate scattering parameter representations of microwave components and antennas. However, circuit simulators, such as SPICE, require conventional equivalent networks with lumped frequency-independent parameters that can be conveniently modeled within CAD-based tools. In view of this, a technique for extracting an equivalent circuit of an implantable antenna is highly desirable in order to ease the design and enhance the reliability of biomedical devices for telemetry. To extract a frequency-independent equivalent circuit of human-body-embedded antennas, a dedicated procedure, based on a heuristic modification of the Foster network synthesis technique, must be developed within this work package. Elements that are resistant to model ohmic and radiation losses should be introduced. In particular, the input impedance of antennas is modeled by means of a suitable equivalent network consisting of R-L-C oscillating bipoles relevant to the resonant processes occurring in the structure, which can be usefully tuned to achieve specific signal- and wave-processing capabilities.

5.5 Experimental Facilities for Small Optoelectronic Sensor

A research platform should support all collaborators in the design, fabrication, and testing of innovative small optoelectronic sensors. A platform based on micro- or nanotechnology should provide facilities about the design, fabrication, characterization, and implementation of innovative materials or technologies. Main research units focus on a particular topic, developing and constructing their facilities only in the examined in order to quickly produce scientific results. Initially it is preferable to collaborate with other specialized research units using their background and experience to create synergistic research units. The main goal of a research platform is also to create an international network in order to propose a project for a small/medium/large enterprise involving collaborations and synergistic works (important aspects for the acceptance of a project).

The main facilities to create are the following:

- *Spectroscopy Laboratory*
  The spectroscopy laboratory can be used to measure optical sensors, optical waveguides, optical fiber communication systems, or optoelectronic systems.
- *Clean Room*
  A clean room is basically used for MEMS and micro-/nanotechnology. The clean room is where devices are fabricated by wet etching processes or by lithography. A clean room contains the following facilities:

  (a) Microscopes
  (b) Scriber (for cutting of wafers)
  (c) Profilometer (to check film thicknesses)
  (d) Mask aligner (lithography)

- *Laser and Light Sources*
  The laser sources can be used for micro-/nanopatterning and for laser ablation techniques for the preparation of metallic nanoparticles. Other optical sources (such as broad lamp tungsten lamp, or IR light source) can be used for different working wavelengths of opto devices.
- *Electrical Device Laboratory*
  The machines included are LCR meter workstations, micro welding machine, digital oscilloscopes, optoelectronic components, I(V) reader probe station, and others.
- *Biology Laboratory*
  In the biology laboratory are generated functionalized surfaces for biosensors, cell cultures, and facilities based on fluorescence for the imaging process.
- *Nanocomposite Fabrication Technology*
  A laboratory that uses smart materials: polymers will be processed for polymerization, and micro-/nanofillers will be generated by chemical techniques.

- *Image Processing*
  Scanning electron microscope (SEM) images are used mainly for MEMS technology to check the technological intermediate results. It is preferable for microscale devices.
  Transmission electron microscope (TEM) approaches are typically used for nanostructures in a solution: TEM samples are typically prepared by dropping a dilute solution of nanoparticles in distilled water and acetone on proper grids.
  Atomic force microscopy (AFM) is used to characterize NMs and nanostructures. An error image provides details about the small signal variation of the topographic surface; the phase information can indicate variations in the material composition; finally, 3D information points out height amplitude variations of the surface. Other AFM tools include:

  (a) Contact mode
  (b) Tapping mode (or intermittent mode, AC mode, noncontact mode)
  (c) Phase imaging
  (d) Force spectroscopy
  (e) Lateral force microscopy
  (f) Force modulation
  (g) Piezo force microscopy
  (h) Conductive AFM with I-V spectroscopy, in both constant current and constant voltage mode
  (i) Electric force microscopy (EFM)
  (j) Surface potential (or Kelvin probe) microscopy
  (k) Scanning tunneling microscopy (STM)
  (l) Electrochemical AFM
  (m) Magnetic force microscopy (MFM)

- *Nanoindentation*
  A nanoindenter can be used to record small loads and displacements with high accuracy and precision. It is an important tool in nanomechanics.
- *Dynamical Mechanical Analyzer (DMA)*
  A DMA can be used to characterize the elastomeric mechanical reactivity of a nanocomposite sample). Elastomeric deformation indicates the slow deformation of a used sample measured under constant stress (the deformation is monitored as a function of time). The sample is then released to an unstressed state and the sample deformation recovery is monitored.
- *Tribometer (Tribology Characterization)*
  This instrument measures friction forces, friction coefficients as a function of disk velocity. This tool can be used for the characterization of shear forces for tactile sensors that use nanocomposite tips.
- *Electrospinning*
  An electrospinning tool is used to generate, by an electrical charge, electrospun nanofibers starting from a liquid solution.

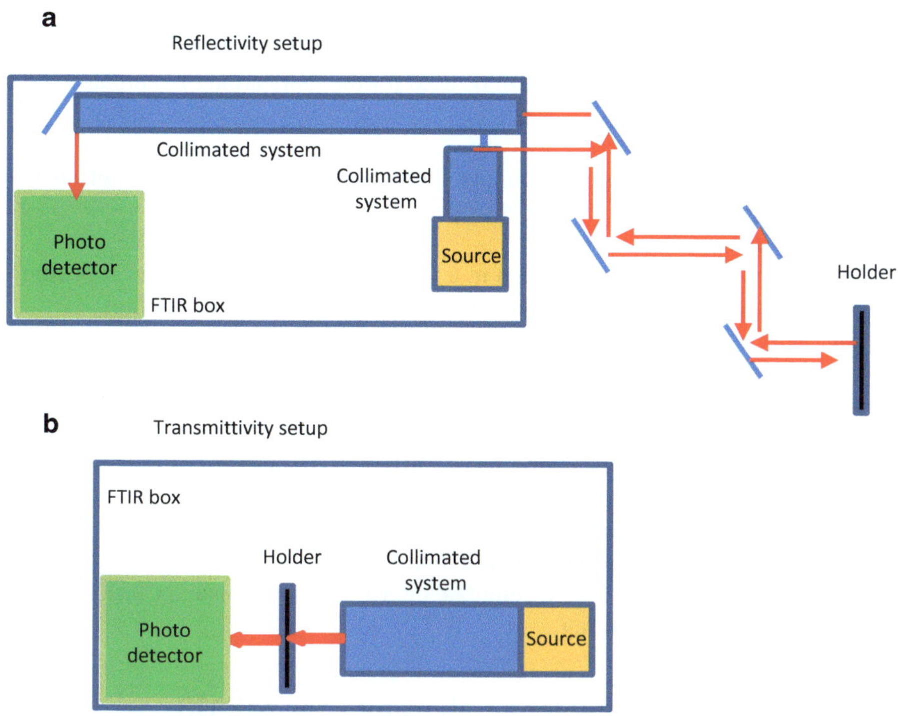

Fig. 5.4 FTIR configurations: (**a**) schematic configuration of reflectivity setup; (**b**) schematic configuration of transmissivity setup

- *Chemical/Physical Characterization*

  Fourier transform infrared spectroscopy (FTIR) can be used to obtain an infrared spectrum of absorption or emission of a NM. Figure 5.4 shows some FTIR configurations. The ultraviolet–visible spectrophotometry (UV–Vis or UV/Vis) characterizes the absorption spectroscopy or reflectance spectroscopy in the ultraviolet–visible spectral region. This tool can be used to characterize the plasmonic resonance of NMs.

  X-ray photoelectron spectroscopy (XPS) is a spectroscopic technique that measures the elemental composition and electronic state of all the elements that exist within a nanocomposite sample. This tool can provide information about the presence of pure metallic nanoparticles proving the nanoparticle generation process.

- *RF Sputtering*

  A sputtering process consists of the erosion of a material from a target (atomic scale) and the formation of a thin layer of the extracted material on a proper substrate. The RF sputtering technique can be used for the fabrication of planar waveguides and for MEMS technology.

- *Electron Beam Lithography*
  During the electron beam lithography process a beam of electrons is emitted in a patterned way. This technique can be used for the fabrication of photonic crystal sensors.
- *Reactive Ion Etching*
  Reactive ion etching is a technique that uses chemically reactive plasma to remove material deposited on substrates.
- *Thermal Evaporate*
  The source material is evaporated in a vacuum, allowing a thin film deposition. Together with RF sputtering, this tool is useful for MEMS technology.
- *Thermogravimetric Analysis and Differential Scanning Calorimetry (Material Characterization)*
  Thermogravimetric analysis (TGA) characterizes the physical and chemical properties of materials as a function of temperature and time. This approach can be used to study crystal structures. Differential scanning calorimetry (DSC) is a thermoanalytical approach and can be used to study what happens to polymers when they are heated.
- *Computational Clusters (Design and Simulations)*
  Numerical tools can be used to simulate micro- and nanosensors. Clusters can be used for parallel computing.

5.6 Creating a Technology Laboratory for MEMS and NEMS

The main goal in creating an educational laboratory in nanotechnology is to include basic chemical approaches a basic instrumentations in order to plan training courses oriented toward basic measurements of nanosensors, MEMS, and NEMS. A basic laboratory could be organized into different sections based on academic level. Instrumentations and laboratory tools can be organized in a modular way for students, Ph.D. students, and young researchers.

An educational laboratory can have a first part regarding the design and the modeling: approaches to the design of innovative devices would be organized into different areas such as biosensors, wireless sensors, micro-/nanoantennas, RF/microwave sensors, sensors for robotics, and nanoelectronics. Experimentation must include the characterization of chemically created bulk/type innovative materials (e.g., semiconductors, polymers, nanocomposite materials). Basic academic experimentation involves the following issues:

1. Current voltage characteristics of bulk-type samples
2. Measurements of scattering parameters for micro-/millimeter antennas and RF/MEMS basic sensors (by means of a network analyzer)
3. Measurements of optical spectra and absorptions of bulk-type materials and of basic optical sensors (by means of an optical source combined with an optical spectrum analyzer)

4. Design and modeling by standard electromagnetic simulators (PC with FDTD, FEM numerical tools, and others)
5. Fabrications of small and basic electronic circuits (basic instrumentation for the fabrication of electronic circuits)
6. Postprocessing approaches to experimental measurements (PC tools)
7. Study of basic packaging layouts
8. Training on the correct use of facilities in an advanced laboratory
9. Chemical fabrication (basic principles) of materials with nano inclusions (NMs and NEMS) and thin film technology (MEMS)

The main goal in the creation of a nanotechnology laboratory, specifically for advanced research, is to create a research platform capable of fabricating and measuring devices on the micro/nano scale.

Advanced laboratories can be geared toward the process of fabricating and testing micro-/nanosensors including wireless sensors. The research activities would be as follows:

1. Design and modeling: design of numerical tools and codes
2. Piezoelectric MEMS: fabrication, testing process, measurements
   - RF-MEMS: fabrication, testing process, measurements
   - Microantennas: fabrication, testing process, measurements
   - Opto-MEMS: fabrication, testing process, measurements
   - Bio-MEMS: fabrication, testing process, measurements

A typical advanced nanotechnology laboratory would include the facilities listed in the previous section.

Acknowledgments The author gratefully acknowledges the help of Diego Caratelli.

References

Ben Hmida, G., Lay-Ekuakille, A., Kachouri, A., Ggariani, H., Trotta, A.: Extracting electrical power from human body for supplying neuronal recording system. Int. J. Smart Sens. Intel. Syst. **2**(2), 229–245 (2009)

Guoyai, J., Tintin, Q., Zhigang, S., Xiaoyan, Z.: A circuit simulating method for heat transfer mechanics in human body. IEEE Proc. EMBS, 5274–5276 (2004)

Jiang, G.T., Shang, Z.G., Umekawa, N.: Study on clinical diagnosis method of diabetes thermography. J. Jap. Soc. Themol. Technol. **1**, 30–33 (2011)

Kumar, D., Sharma, R.: Advances in conductive materials. Eur. Polym. J. **34**(8), 1053–1060 (1998)

Massaro, A., Spano, F., Lay-Ekuakille, A., Cazzato, P., Cingolani, R., Athanassiou, A.: Design and characterization of nanocomposite pressure sensor implemented in tactile robotic system. IEEE Trans. Instr. Meas. **60**, 2967–2975 (2011)

Simon, H.J., Mitchell, D.E., Watson, J.G.: Surface plasmons in silver films a novel under graduate experiment. Am. J. Phys. **43**(7), 630–636 (1975)

Chapter 6
Instrumentation for Measurement Procedures

Abstract Building electro-optical instrumentation to take measurements is not easy because of the various factors can introduce inaccuracies, errors, and bias. It is necessary to know the common parameters that should be taken into account when the decision is made to build an electro-optical system. This chapter illustrates some interesting applications in the area of using led for public lighting and photoacoustic spectroscopy.

6.1 Measurement Issues and Its Impacts

When a decision is made to proceed with a feasibility of a measurement, there must be a balance between theory and experiment. Ideally, theory shows that a measurement is possible and how big the signal should be; experiment shows whether the theory is right (often leading to improved theories) and that the instrumentation can be built. Combining theory and data, by demonstrating good agreement, shows that the measurement is at least possible; the actual instrument design is less important because there are usually a few different good ways to go about it.

The quality of the signal to be measured is an essential aspect of measurement, especially in optics. This quality affects the right measure and to be taken into consideration and that must be closer to what we need to know. In optics, many issues are important such as, for example, phase measurements, dynamic range, fringes, and photon quantities. It is important to establish whether a measurement is phase insensitive since the use of filtering can constrain the phase and amplitude. A good wide-range position of measurements can be taken using an interferometer and a tunable light source. Fringes and speckle corrupt all kinds of measurement by introducing disturbances and interference that can be reduced or positively exploited to measure surface geometry using mirrors and interferometry. Another issue is preserving photons. The use of a staring detector, for example, allows one to measure all grass at once rather than scanning one blade at a time. To increase the

A. Lay-Ekuakille, *Optical Waveguiding and Applied Photonics*, Lecture Notes in Nanoscale Science and Technology 10, DOI 10.1007/978-1-4614-5959-0_6,

quantity of photons, and hence the optical signal, it is important to adjust the system sensitivity, time-delay integration, slitless spectroscopy, corralling photons, etc.

6.2 Common Standards for Measurement and Instrumentation Parameters

It is very rare to encounter a compact and complete optical instrument for measurement. In general optical instruments are semicustomized so that the operator can add or remove parts to enhance the system. Updating necessitates that specific requirements be known to avoid problems and malfunctions. Accuracy is one of the most important parameters. We detail in Table 6.1 a list of parameters that are necessary for building a semicustomized optical system.

We can use design parameters as elements for semicustomized systems. As an example we present a lens design for led applications to be used in public lighting and other areas. The list of specifications is very long, but it is important to know some of the main characteristics: aberration origin, spherical aberration, relationship between aspheric surfaces and aberration, and reciprocal aberration.

We use a led as LUXEON K2 DS60 (Philips 2012) with a led matrix and with a beam angular aperture as 140° × 60°. Figure 6.1 illustrates the mechanical characteristics and the angular aperture. Instead, Fig. 6.2 depicts the design of the led matrix that will host the lens to be designed and constructed.

Table 6.1 Acceptable limits for common design parameters (Hobbs 2000)

Parameter	Characteristics
Bandwidth	10 Hz≤BW≤50 GHz
Pel rates	≤1 G pixels/s for scanning systems
Motor speed	≤5,000 rpm
Mirror flatness	1/15 at 655 nm
Etalon fringes versus bandwidth and spatial coherence	0.5 % P-P, slope 15 %/GHz collimated
Polygon alignment	15 arc s
Mechanical backlash	5 μm
Eye-safe lasers	1 mW visible, 100 μW IR
Diffraction limit	λ/5 rms wavefront error → 0.8 strchl ratio
Diode laser feedback tolerance	3×10^{-7}
Diode laser current tunability	10 cm^{-1}
Laser pointing stability	5 arc s
Diode laser lifetime	100,000 h at 25 °C
Electronic component stability	100 ppm/°C
Temperature coefficient of index	10^{-5} glass, 10^{-4} plastic

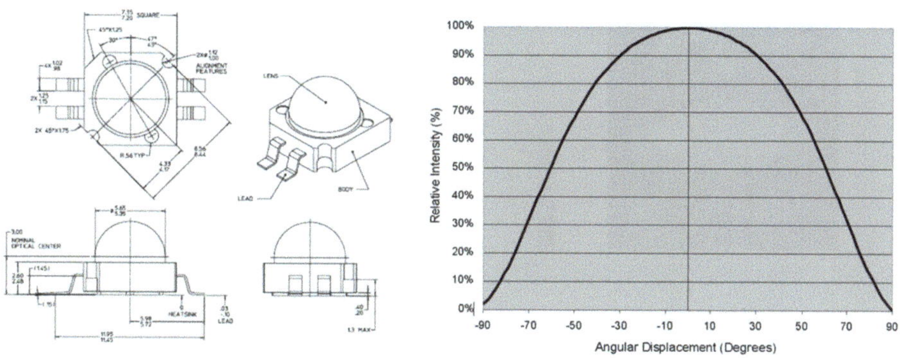

Fig. 6.1 Luxeon K2 mechanical characteristics and angular displacement

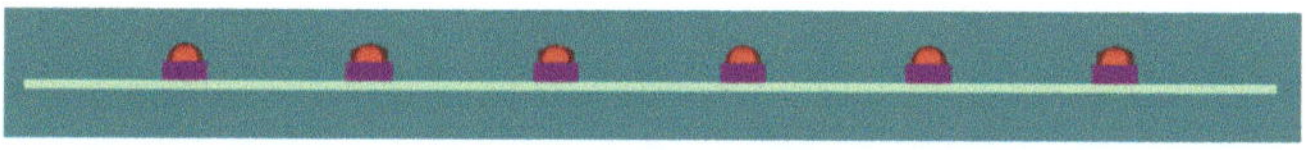

Fig. 6.2 Cross section of matrix of six Leds for Luxeon K2. Dimensions are 200 × 145.50 mm

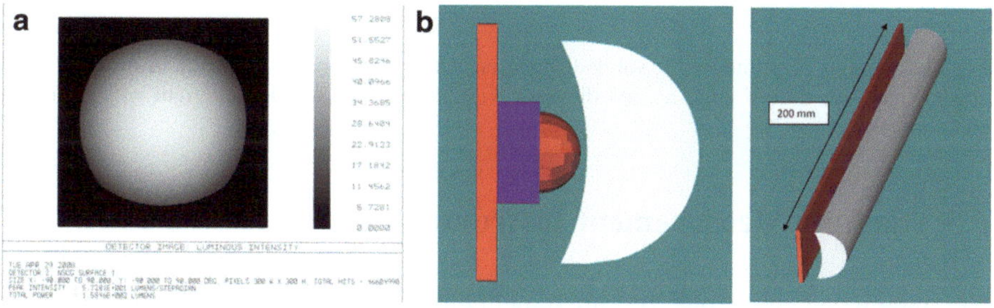

Fig. 6.3 Lens luminous intensity (**a**) and simulated lens for one section of matrix of Fig. 6.2(**b**)

Zemax software (Zemax 2008) is used to model the lens; Fig. 6.3a shows the trend of luminous intensity as a function of irradiance. The light flux is 160 lm with respect to the theoretical 170 lm. Now we can see the actual shape and volume of the lens for lighting, as shown in Fig. 6.3b.

The results of simulation using Zemax design software are illustrated in Fig. 6.4, which shows evidence of a good design that allows one to use the lens for led-based public lighting. It is now possible to build the lens for the fixture for testing in order to retrieve the photometric solid, as indicated in Fig. 6.5a. The outputs of lens testing on a fixture pole are shown in Fig. 6.5b, c.

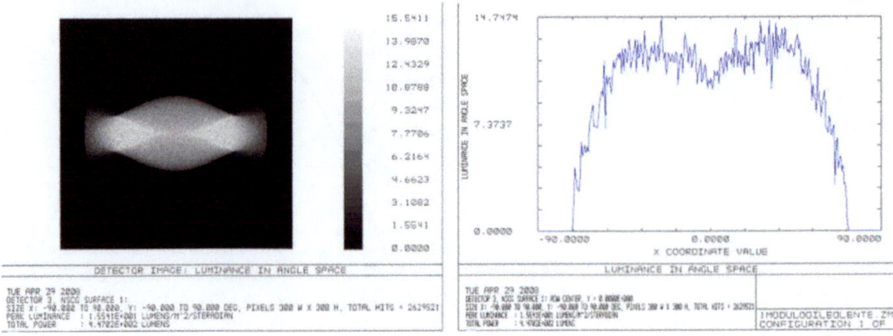

Fig. 6.4 Luminance versus irradiance (*left*) and angular distribution versus luminance along *x*-axis

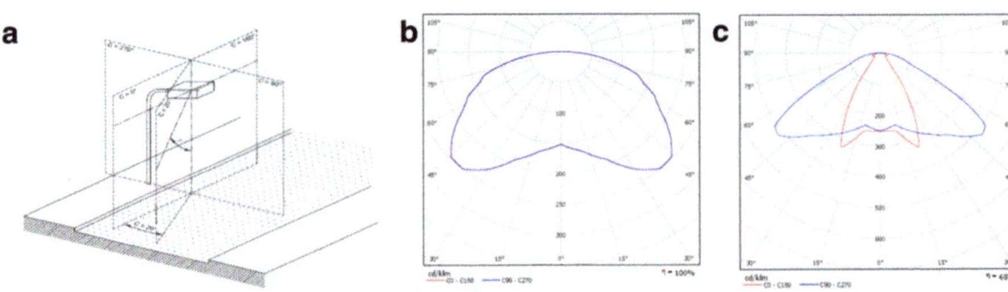

Fig. 6.5 Image of photometric solid disposition for testing lenses included in array of leds and mounted on fixture pole (**a**), led diagram without lens (**b**), and led diagram with designed and constructed lens (**c**)

6.3 Sample Measurement Instrumentation

An interesting example of instrumentation design and construction is a photoacoustic system dedicated to indoor gas detection, and in particular for surgery applications. The system uses a light source, but it can be improved using led lamps as the light source. There are various techniques for measuring anesthetic gas concentrations in the air:

1. Direct-reading systems
2. Air sampling with equipment such as adsorbent tubes or cuffs, followed by analysis using gas chromatography or an infrared technique
3. Diffusion sampler and analysis

Considering the variation range of the parameters affecting exposure to anesthetic gases, air monitoring is preferable to other methods of assessing exposure in the workplace. The most commonly employed model for describing the photoacoustic effect in condensed samples was developed in the 1970s by

Rosencwaig and Gersho. Applying Beer's law with radiation intensity I_0 and optical absorption coefficient β yields

$$dI = -\beta I dx \Rightarrow I = I_0 e^{-\beta x}. \tag{6.1}$$

Suppose the incident radiation is modulated with frequency ω. Then the incident intensity is given by

$$I' = \frac{1}{2} I(1 - \cos\omega t). \tag{6.2}$$

Both the sample and the gas must satisfy the heat-diffusion equation, which for the case of the sample is given by

$$\frac{\partial^2 T}{\partial x^2} = \frac{1}{\alpha}\frac{\partial T}{\partial t} - \frac{\beta \sigma_{ri} I'}{k} = \frac{1}{\alpha}\frac{\partial T}{\partial t} - \frac{\beta \sigma_{ri} I_0}{2k} e^{-\beta x}(1 - \cos\omega t), \tag{6.3}$$

where σ_{ri} is the probability of a radiationless transition, and the thermal diffusivity $\alpha = \frac{k}{\rho c}$,where k is the thermal conductivity of the sample, ρ is the density, and c is the specific heat. Rosencwaig and Gersho came up with the following equation for the temperature in the surrounding medium as a function of both position and time:

$$T^{ac}(x,t) = e^{-ax}[\theta_1 \cos(\omega t + ax) - \theta_2 \sin(\omega t + ax)], \tag{6.4}$$

where the complex temperature amplitude $\theta = \theta_1 + i\theta_2$, and the thermal diffusion coefficient $a = \sqrt{\frac{\omega}{2\alpha}}$.

The equation describes a periodic temperature wave that propagates through the medium surrounding the sample. The temperature fluctuation described by this equation is the cause of the pressure waves that are detected. Since the e^{-ax} causes the wave to decay away from the sample, the sensor should be located within the thermal diffusion length $\mu = \frac{1}{a} = \sqrt{\frac{2\alpha}{\omega}}$ in order to maximize the strength of the acoustic signal. Surprisingly, modern photoacoustic analysis has not deviated far from Bell's original "chopped light" setup, apart from the introduction of lasers and microphones. A typical photoacoustic experiment (Fig. 6.6) consists of a laser incident (Lüscher 1984) upon a chamber, which is modulated with a physical "chopper" or some other method of precisely pulsing light. The acoustic signal is measured with either a microphone or piezosensor depending on the frequency. Frequencies of applied light from microwaves to X-rays can be used, but lasers or xenon lamps tend to be the most common. Sensitivity is of great importance in any type of photoacoustic measurement, so a chamber containing a sample for analysis must be well insulated from outside noise and vibration. The photoacoustic effect due to surrounding materials such as the chamber itself must also be minimized – even the chamber itself will demonstrate some amount of photoacoustic emission, and this effect must be taken into account by the experimenter. Modern photoacoustic

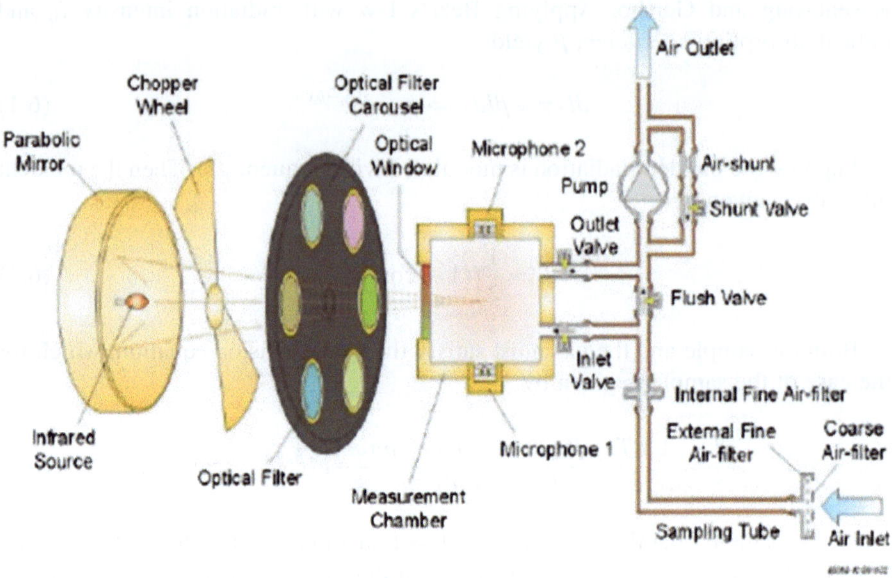

Fig. 6.6 Photoacoustic instrumentation principle

experiments commonly employ chambers with cylindrical or spherical symmetry in order to take advantage of acoustical resonances (Hess and Fiedler 1989). In this case, a standing wave is formed by tuning the applied laser to the resonance frequency of the chamber. This resonance amplifies the sound signal and allows it to be more easily detected.

Figure 6.7 illustrates a commercial photoacoustic instrumentation used for anesthetic gas measurements, while Fig. 6.7b, c shows the electronics from which, using clamps, signals are taken for analysis in order to design a new sensor control board according to the overall scheme illustrated in Fig. 6.8.

As indicated, blocks relative to sensor units and impedance adaptation buffers are included in the scheme for completeness, but they are not really included in the board. After successive treatments by means of LabVIEW virtual instrumentation, a board is designed, as demonstrated in Fig. 6.9.

The new sensor control board was made as simple as possible, so that the CPU has no constraints in processing measured data. This makes it possible to reduce production costs for this kind of instrumentation. Another question has been addressed concerning the design of a new chamber capable of measuring indoor gas in surgery rooms (Rosencwaig 1980). The major advantage of such a chamber (Fig. 6.10) consists in a double light source, instead of one, according to commonly used photoacoustic apparatus, which increases the accuracy of measurements by allowing a double analysis of the same gas. This also increases the speed of analysis in acquiring a gas when a multisampling facility is used.

To control and move correctly the two stepper motors, namely, one for the chopper wheel and the other for the optical filter carousel, an appropriate electric

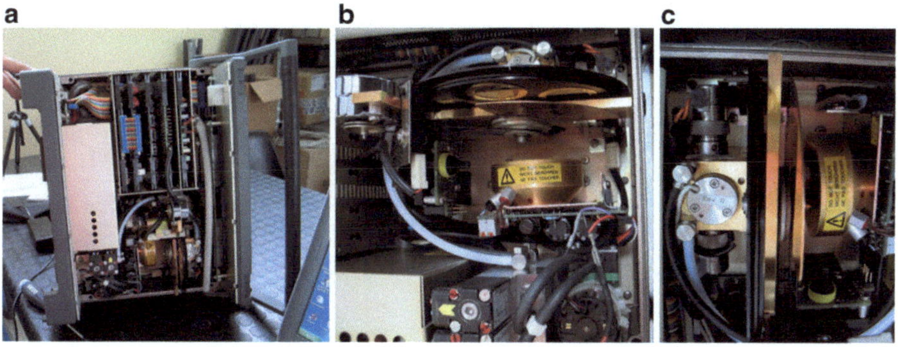

Fig. 6.7 Commercial photoacoustic instrumentation from which signals are extracted: overview (**a**, *left*), filter and chamber (**b**, *center*), and chamber top view (**c**, *right*)

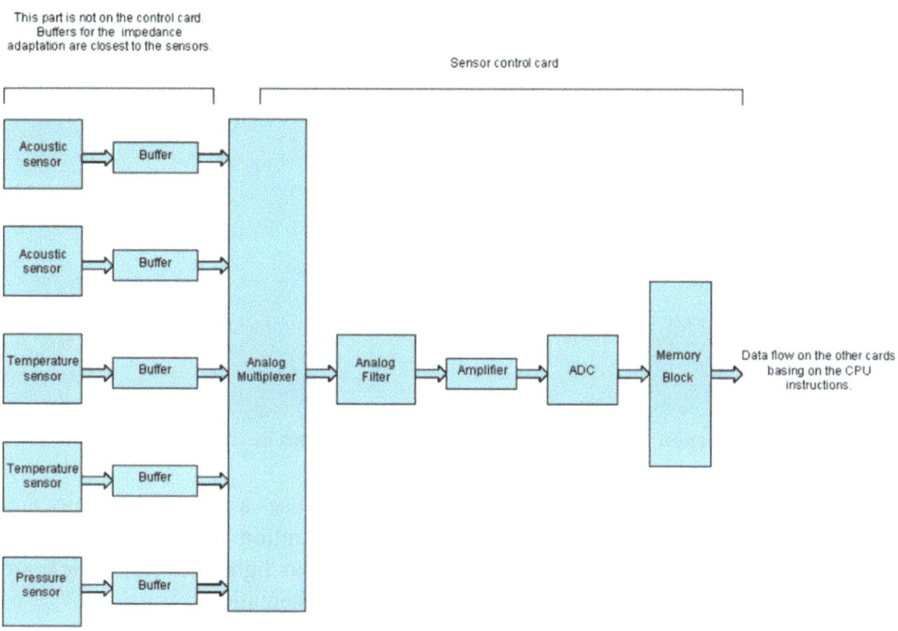

Fig. 6.8 Block diagram of sensory system for photoacoustic instrumentation

circuit was made and tested for this purpose, as indicated in Fig. 6.11. Finally, it is possible to test the chamber for an actual case. Since many perfumes (Les phthalates 2005) contain solvents (e.g., phthalates), which are similar to those recovered in surgery rooms, different sessions of measurements have been carried where the perfume spectrum is converted into voltages (Fig. 6.12). Various spectra converted into voltages agree with the perfume composition of different producers (Fig. 6.13).

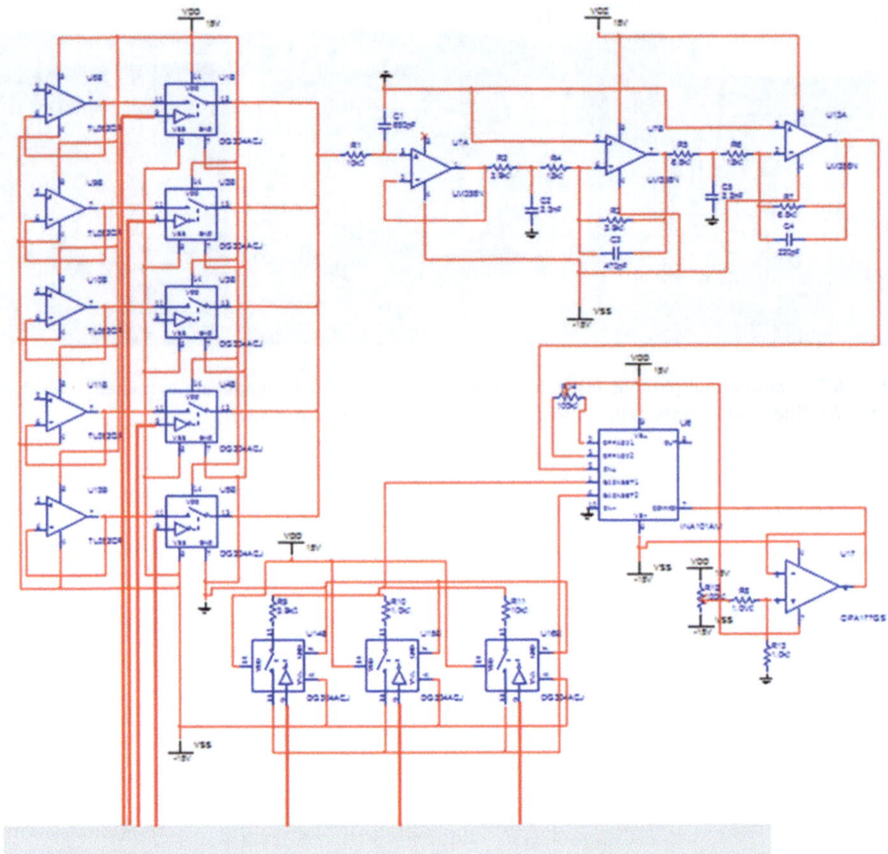

Fig. 6.9 Designed sensor control board. The strip below indicates the control unit

Recall that the command signal of a stepper motor uses a PWM technique, and the system has a second-order behavior. Special attention must be paid to the transfer function G(s) because of the presence of two light sources, which are controlled by the electronic control indicated in the schematic.

Even if, based on theoretical considerations, one could predict the second-order-based behavior of a system, in practical terms, this is impossible; the system response can be anomalous and disturbed by the simultaneous power feeding of the two components NJM1 and NJM2 (Fig. 6.11, right). An incorrect synchronization between the instruction dedicated to the PIC (microcontroller) and putting into effect the NJM1 (and NJM2) can generate mismatching that may provoke an unexpected trend after the maximum point is reached by the actual transfer function and, consequently, undesired oscillations. Stepper motors can often exhibit a phenomenon referred to as resonance at certain step rates. This can be seen as a sudden loss or drop in torque at certain speeds, which can result in missed steps or loss of synchronization. It occurs when the input step pulse rate coincides with the natural oscillation frequency of the rotor.

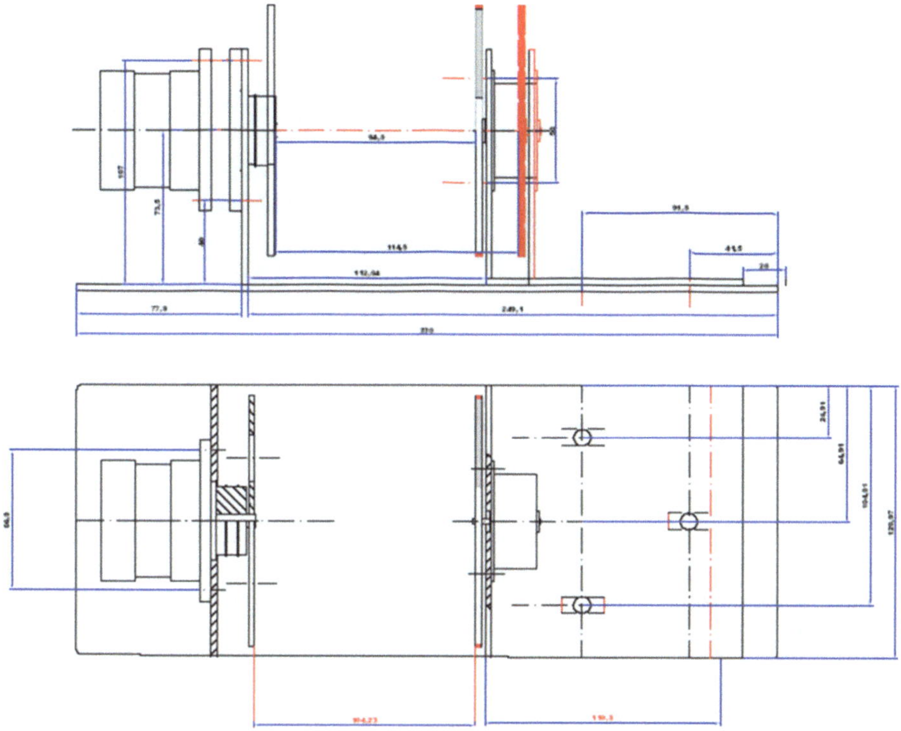

Fig. 6.10 Proposed photoacoustic chamber: side view (*upper*) and top view (*bottom*)

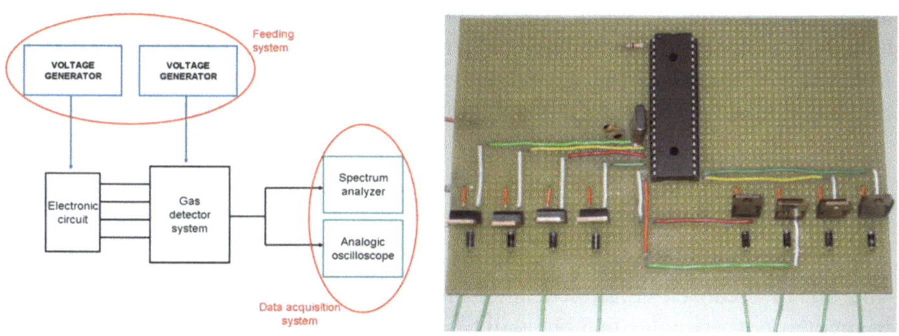

Fig. 6.11 Block scheme of electronic control (*left*) and stepper motor board (*right*)

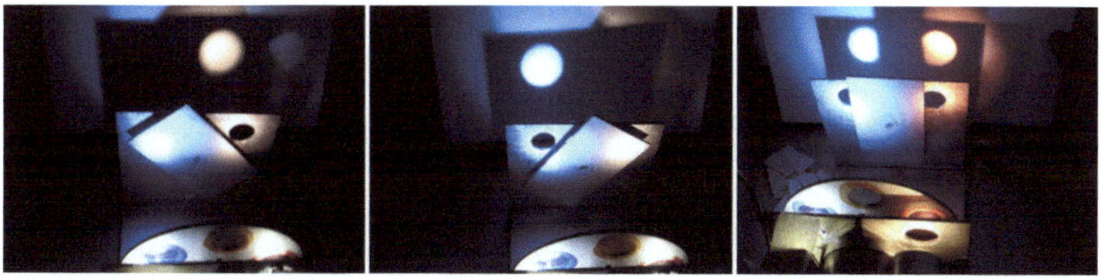

Fig. 6.12 Optical testing versus carousel – chopper positions

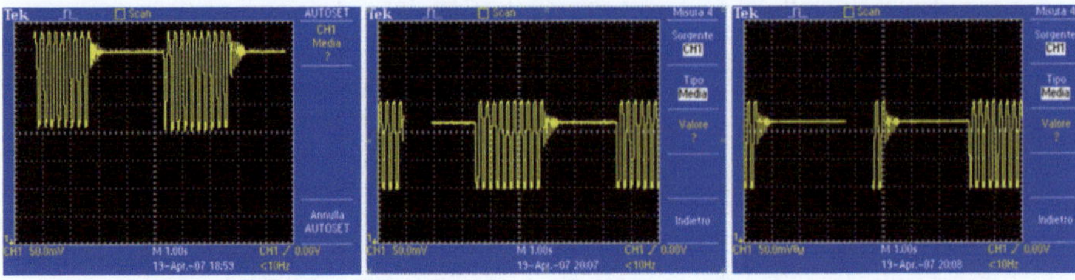

Fig. 6.13 Spectra recovered from optical testing using solvents included in different vials

6.4 Experimental Design: Errors, Accuracy, and Repeatability

In the design of optical systems, many errors occur that lead to lower accuracy and reduce experimental repeatability. Different solutions can be adopted, keeping in mind that the most important thing is to reduce background fluctuations that greatly affect optical system errors.

6.4.1 Beam-Pointing Stabilization

Active optics beam-pointing stabilization, as illustrated in the previous examples, are based on shining a sample of beam into a quad cell and tipping mirror to keep it still. In a narrow bandwidth they can do a very good job. One limitation is the interaction of angular and position wobbles; focusing the beam on the quad cell with a lens will make the system ignore wobbles in position but correct pointing errors.

6.4.2 Beam Intensity Stabilization

The use of feedback-controlled attenuators that regulate the intensity of a laser beam does not guarantee that one will be able to keep the beam constant because of continuous variations and instability. There is no way to stabilize a beam down close to the shot noise in this way without wasting a whole lot of it; a 50 % loss will ideally get you to 3 dB above the noise.

6.4.3 Photocurrent Stabilization

In the case of bright-field measurements and characterization, it is very hard to stabilize a beam. One way to do this would be by using differential measurements of the photocurrent, where a comparison current derived from a sample of the source beam is subtracted or divided out. Since both beams will see the same fractional noise, this will leave a quiet result.

For the aforementioned reasons, we assert that systematic and random errors during experimentation or common use of optical systems produce uncertainty. The degree of inaccuracy or the total measurement error is the difference between the measured value and the true value. The total error is the sum of the systematic (or bias) error and the random (or precision) error. The systematic error is a fixed or constant component of the total error and is sometimes referred to simply as the bias. The random component of the total error is sometimes called the repeatability, repeatability error, or precision error.

Acknowledgments The author gratefully acknowledges the technical support of Sud Segnal Srl and Revi Srl companies.

References

Hess, P., Fiedler, M.: Laser excitation of acoustic modes in cylindrical and spherical resonators: theory and applications. In: Hess, P. (ed.) Topics in Current Physics: photoacoustic, Photothermal and Photochemical Processes in Gases, p. 85. Springer, Berlin/Heidelberg (1989)

Hobbs, P.C.D.: Building Electro-Optical Systems Making It All Work. Wiley, New York (2000)

Les phtalates en cosmetology, VIGILANCES- Bullettin de l'Agence française de sécurité sanitaire des produits de santé, no. 28 (2005)

Lüscher, E.: Photoacoustic effect in condensed matter—historical development. In: Lüscher, E., et al. (eds.) Photoacoustic Effect: principles and Applications, p. 1. Friedr. Vieweg & Sohn, Braunschweig (1984)

Philips: http://www.philipslumileds.com/products/luxeon-rebel (2012)

Rosencwaig, A.: Photoacoustics and Photoacoustic Spectroscopy, p. 12. Wiley, New York (1980)

Zemax: www.zemax.com (2008)

Chapter 7
Electronic Measurements and Signal Processing

Abstract Electronic measurements and related signal processing are basic topics for understanding information acquisition, transmission, and its further use. In optical systems these topics are mostly important because of the use of light and its eventual interferences. This chapter illustrates the basic chains used in measurement procedures, methods of measurement, and issues regarding signal processing.

7.1 Analog and Digital Chains of Measurement

The role of measurement chains is to collect information from a quantity under measurement (QUM) by means of specific sensors (or transducers). The chains are a summary of measurement architecture in any field, even for optical applications. An analog chain (Dally et al. 1993), as illustrated in Fig. 7.1, starts with the transformation of a physical signal into an electrical one using a sensor (or transducer), a SCB is required for conditioning the signal in terms of amplification or attenuation. Then a filter is needed and the signal can be used immediately or recorded for further utilization. In a digital chain (Nawrocki 2005), instead, beyond the previous blocks, it is necessary to add an anti-aliasing filter before converting the signal by means of an ADC. The signal can be directly used by a computer, and it also follows a feedback block for controlling the error between the initial signal acquired at the sensor interface and that processed by all blocks, as depicted in Fig. 7.2. Both chains are basically the point from which one can construct one's own architecture for measuring and acquiring any kind of signal. They are used for either engineering analysis or process control. Both chains must respect some imposed operating mode parameters in order to be useful for the operator. These parameters are as follows (Northrop 2005): interval of measurement, precision, resolution, quickness, and immunity from interferences.

A. Lay-Ekuakille, *Optical Waveguiding and Applied Photonics*, Lecture Notes in Nanoscale Science and Technology 10, DOI 10.1007/978-1-4614-5959-0_7,

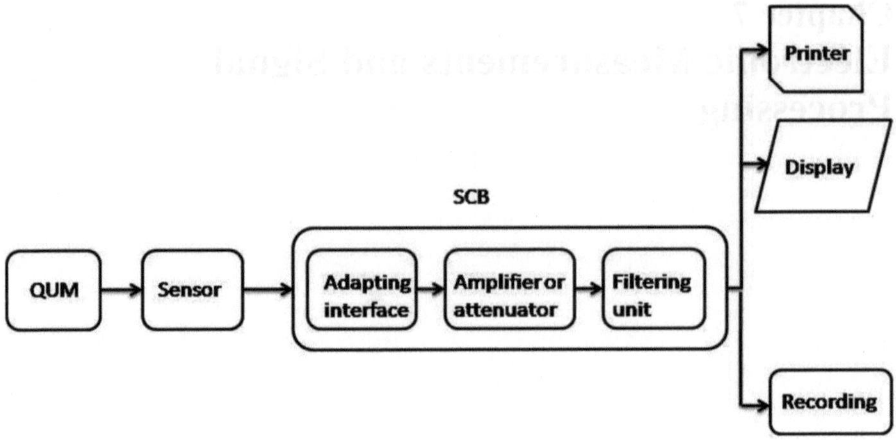

Fig. 7.1 Analog chain of measurement: *QUM* quantity under measurement, *SCB* system conditioning block

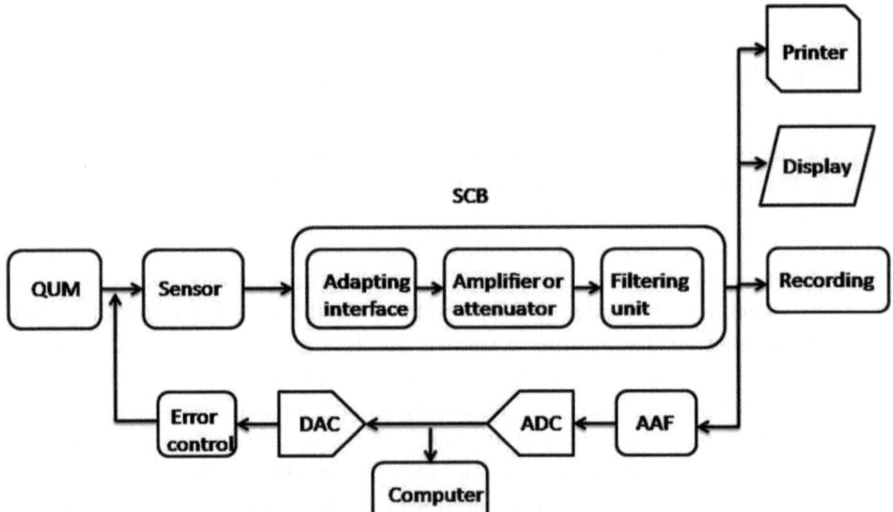

Fig. 7.2 Digital chain of measurement: *QUM* quantity under measurement, *SCB* system conditioning block, *AAF* anti-aliasing filter

The aforementioned chains can also be used when it is appropriate to acquire data from multiple measurands, each one having its proper channel (e.g., sensor, conditioning block). A multiplexer allows one to select a certain channel so that the specific signal is received by the appropriate downstream devices.

7.2 Types of Measurement

Many kinds of measurement can be performed in the optical field related to waveguides and nanocomposites. Spectroscopy—infrared, infrared Fourier spectroscopy, dark-field, laser scanning, and differential procedures—is mostly used in this area of measurement. Infrared (IR) spectroscopy is one of the most common spectroscopic techniques used by chemists. Simply put, it represents an absorption measurement of different IR frequencies by a sample positioned in the path of an IR beam. Examination of the transmitted light reveals how much energy was absorbed at each frequency (or wavelength). IR absorption information is generally presented in the form of a spectrum with the wavelength or wavenumber as the x-axis and the absorption intensity or transmittance as the y-axis. The absorption coefficient here indicated as $k(\nu)$ is defined as the fractional decrease in the flux density at frequency ν per unit path length through the absorbing medium:

$$-\delta I(\nu) = I(\nu)k(\nu)\delta x, \tag{7.1}$$

where $I(\nu)$ is the flux density incident on the layer of thickness $\delta(x)$. The units of k are reciprocal length, or m^{-1}. For a homogenous layer of thickness l this equation can be integrated to give

$$I_l(\nu) = I_0(\nu)e^{-k(\nu)l}. \tag{7.2}$$

The product $k(\nu)l$ defines the optical depth $\tau(\nu)$ at frequency ν. If the absorbing medium is not homogenous, then

$$\tau(\nu) = \int_0^l k(\nu)dx. \tag{7.3}$$

Absorption from an isolated spectral line is spread over a finite frequency range. The absorption coefficient can be easily related to the Einstein coefficient (Okada et al. 2002) by considering the total power absorbed from the incident beam. Integrating equation (7.1) over the lines gives

$$-\Delta I = I(\nu_0)\delta x \int_{line} k(\nu)d\nu \text{ in Wm}^{-2}. \tag{7.4}$$

Absorption is sometimes expressed in terms of the absorption coefficient per absorbing atom, $\alpha(\nu)$. It is worth noting that the relationship between $k(\nu)$ and $I(\nu)$ is logarithmic: unless the absorption is weak, the area under the $k(\nu)$ curve of Fig. 7.3b does not scale with the area defined by the absorption line in Fig. 7.3a.

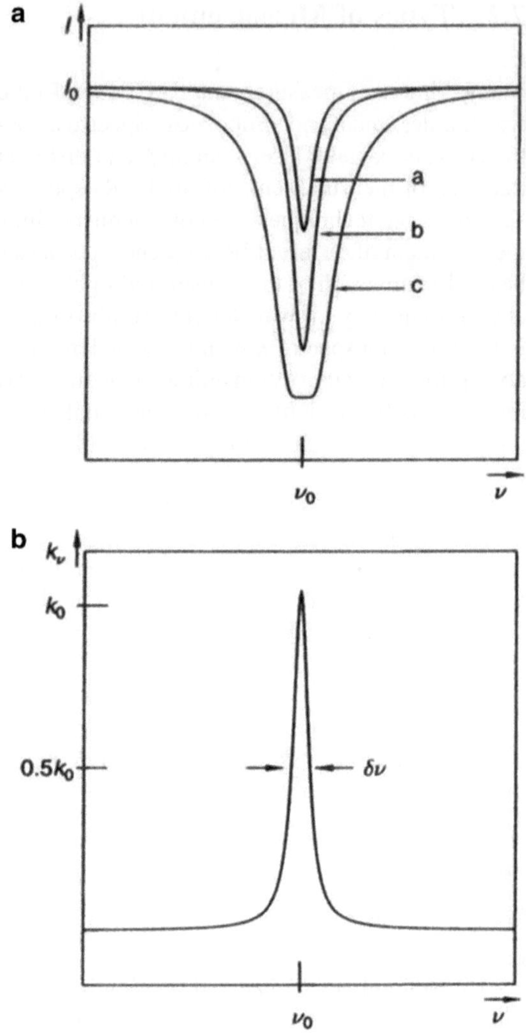

Fig. 7.3 (**a**) Example of absorption line (**b**) and absorption coefficient. The absorption line shows the effect of increasing optical depth on line shape. *a*, *b*, and *c* refer to optically thick regimes (Corney 1977)

The latter, normalized to the incident flux density, is known as the equivalent width and is given by

$$\begin{aligned}\text{Equivalent width} &= \frac{1}{I_0}\int_{line} [I_0 - I(\nu)]\ d\nu = \int_{line} [1 - I(\nu)/I_0]\, d\nu \\ &= \int_{line} \left[1 - e^{-k(\nu)l}\right] d\nu \\ &= \int_{line} [k(\nu)l - k^2(\nu)l^2/2 + \ldots]\, d\nu. \qquad (7.5)\end{aligned}$$

The requirement for the two areas to match is therefore $k(\nu)l << 1$ for all values of ν. Wavenumbers and wavelengths can be interconverted using the following equation:

$$\overline{\nu}\left(cm^{-1}\right) = \frac{1}{\lambda(\mu m)}\ 10^{-4}. \tag{7.6}$$

Absorption spectroscopy is employed to determine the presence of a particular substance in a sample and, in many cases, to quantify the amount of the substance present. Absorbance (A) is the logarithm to base 10 of the reciprocal of the transmittance (T), and transmittance is the ratio of radiant power transmitted by the sample (I) to the radiant power incident on the sample (I_0):

$$A = -\log T;\ \ T = \frac{I}{I_0} \tag{7.7}$$

Absorption and transmission spectra represent equivalent information, and one can be calculated from the other through a mathematical transformation. An absorption spectrum can be quantitatively related to the amount of material present using the Beer–Lambert law, which states that the absorbance of a solution is directly proportional to the concentration of the absorbing species in the solution and the path length. This method is often used in a quantitative way to determine the concentration of an absorbing species in solution:

$$A = -\log_{10}\frac{I}{I_0} = \varepsilon \cdot c \cdot L, \tag{7.8}$$

where A is the measured absorbance, I_0 is the intensity of the incident light at a given wavelength, I is the transmitted intensity, L is the path length through the sample, and c is the concentration of the absorbing species. For each species and wavelength, ε is a constant known as the molar absorptivity or extinction coefficient. An alternative measurement technique to obtain an IR spectrum of absorption is Fourier transform infrared (FTIR) spectroscopy. A FTIR spectrometer collects spectral data in a wide spectral range, and a Fourier transform is required to convert data into a spectrum.

The heart of a FTIR spectrometer (FT-IRS) is an interferometer, which has several basic advantages over a classical dispersive instrument. Some of these advantages are (Griffiths 1975) as follows:

> all source wavelenghts are measured simultaneously in an interferometer, whereas in a dispersive spectrophotometer they are measured successively. A complete spectrum can be collected very rapidly and many scans can be averaged in the time taken for a single scan of a dispersive spectrophotometer; for the same resolution, the energy throughput in an interferometer can be higher than in dispersive spectrophotometer, where it is restricted by the slits. In combination with the multiplex advantage, this leads to one of the most important features of an FT-IR spectrophotometer: the ability to achieve the same signal-to-noise ratio as dispersive instrument in a much shorter time. There are two main limitations to FT-IRS. The first is the cutoff at short wavelengths because of the need for a transmitting

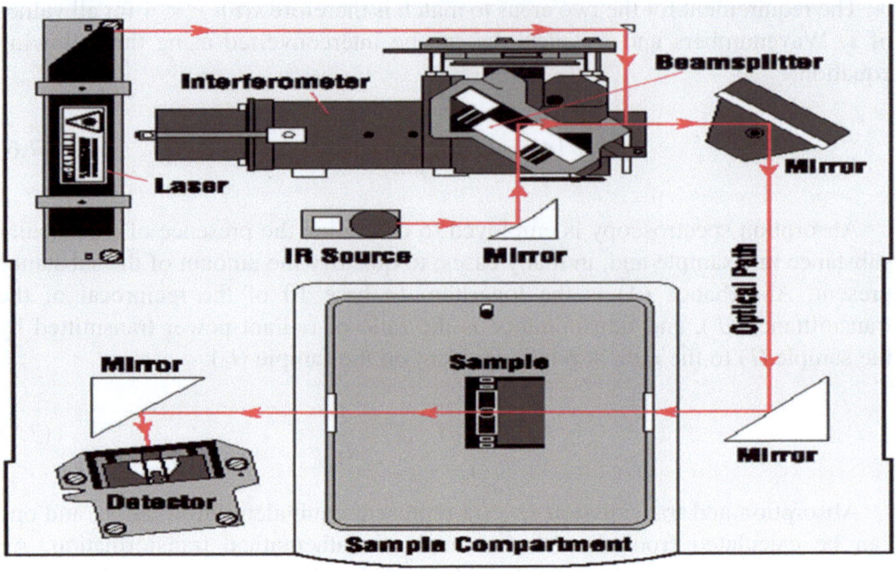

Fig. 7.4 FT-IRS layout example

substrate to act as beamsplitter as shown in Fig. 7.4 (the transmission of fused silica falls sharply below 180 nm) and the stringent optical and mechanical tolerances required. The second limitation is the requirement for a source that remains reasonably constant in intensity over the period of the scan, because fluctuations in intensity translate into noise in the spectrum. In this paper, we take advantage of the speed, power and relative cheapness of modern computers, to build a filtering system and a remote interface (front panel) so that it is possible to guide and control the FT-IRS. In this context, it is suitable to use one instrumentation for different users.

The preceding instrumental layout concerns signal processing of a spectrum, for instance, a pollutant such as SO_2, as illustrated in Fig. 7.5. The spectrum is obtained from the interferogram by Fourier transformation. This process analyzes the interferogram as the sum of a series of sine and cosine waves with discrete frequencies. The range of frequencies used in the Fourier analysis is limited by the way in which the interferogram is stored as a series of discrete points. There must be at least two data points for each cycle if a wavelength is to be recognized correctly (the Nyquist criterion). This means that the interferogram must be measured at intervals equal to half of the shortest wavelength from the source to be measured. Radiation at shorter wavelengths may contribute to the interferogram but will appear at the wrong place in the final spectrum. This effect is known as aliasing. The problem is avoided by filtering our shorter wavelength optically or electronically. Here, we use digital filtering with a least-squares filter. We have used this kind of filter when the aim is to find a filter that minimizes the squared error. The sampling points for measuring the interferogram are derived from the laser signal (Pelkim Elmer 1995). If one data point is recorded for each cycle of the laser signal, then data point separation is equal to a laser wavelength (in a vacuum) of 632.99 nm. The resolution in the final spectrum depends on the maximum optical path difference in the interferogram.

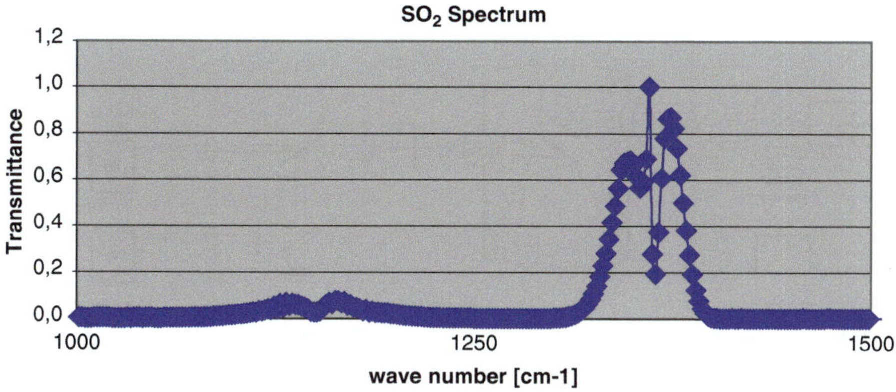

Fig. 7.5 SO_2 spectrum used as sample

The Fourier transformation analyzes the interferogram as the sum of the contributions from individual wavelengths. However, the Fourier transformation is not able to properly distinguish between wavelengths that differ by less than half a cycle. So we have two wave numbers, and $\bar{\nu_1} - \bar{\nu}_2 = \frac{1}{2d_{max}}$, where d_{max} is the maximum optical path difference. A useful approximation is that two spectral features must be at least two such spacings apart to be resolved.

This means that the resolution is approximately the reciprocal of the maximum optical path difference scanned.

A dark-field measurement is a kind of measurement in which the light ideally corresponds to the signal. An example of this is a dark-field microscope (Hu et al. 2008); however, other examples are possible to trace out such as second harmonic generation (SHG), fluorescence, and scintillation. Dark-field measurements are almost always Johnson noise. With laser scanning (Lohmann et al. 2000), one can obtain a flying-spot laser, which is a good way to obtain more signals; nevertheless, it is. It displays a further advantage due to the fact that the detector sees a very brightly illuminated sample point at all times, while the incoherent optical background signal is detected with a 10^{-6} duty cycle. Differential measurements (Lee and Wu 1995) are performed, for example, using dual-beam spectrometers, where one beam passes through the sample and one does not. subtracting two carefully balanced photocurrents ideally cancels the source of excess noise exactly, at least if the noise in the two beams has not been decorrelated by vignetting or etalon fringes.

7.3 Experimental Setup Descriptions, Signal Processing, and Hardware

In this section two interesting examples of optical measurements are detailed: a lidar (light detection and ranging) system and optical measurements for automotive applications. All single-ended lidar systems consist of a laser and telescope whose

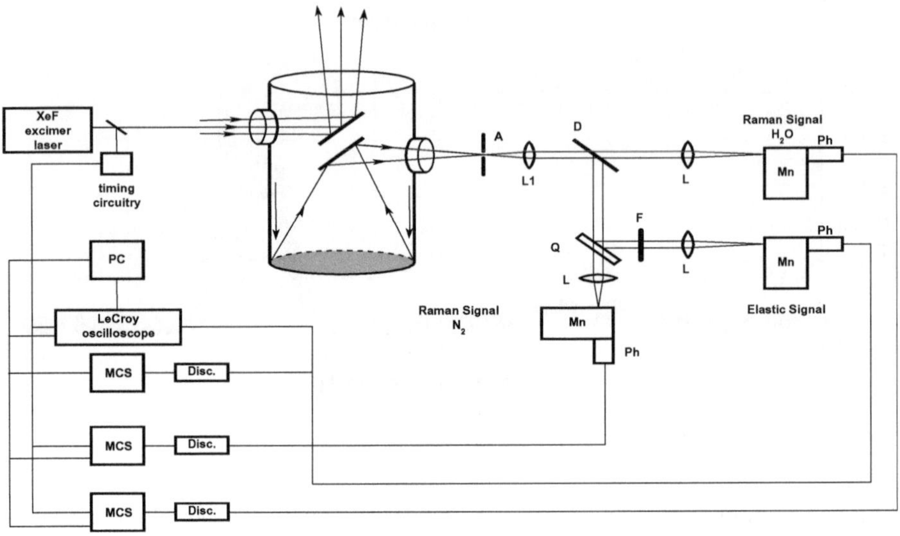

Fig. 7.6 Experimental lidar general scheme

optical axes are aligned parallel such that the telescope field of view includes the laser beam as it propagates through the atmosphere. The differences in the lidar systems mentioned previously lie in the selection of the laser wavelength, the receiver wavelength, and data analysis and interpretation. In the most basic system, referred to simply as lidar, the detector wavelength is matched to the laser wavelength. As the laser radiation propagates through the atmosphere, it interacts with aerosols and molecules. The principal interaction of interest is elastic scattering by aerosols. Some of the scattered radiation is collected by telescope and detected by a sensitive photomultiplier. The detected signal is recorded as a function of time to provide a range-resolved measure of atmospheric scattering. Analysis of the recorded data is then performed to provide an indication of aerosol distribution in both the troposphere and stratosphere. In Raman lidar, which we use, the receiver is made sensitive, through the use of a spectrometer or interference filter, to Raman-shifted wavelengths. Most molecules scatter electromagnetic radiation not only at the excitation wavelength but also at specific shifted wavelengths by the phenomenon referred to as Raman scattering. The magnitude of the shift is unique to the scattering molecule, and the intensity of the Raman band is proportional to the scattering molecule's concentration. Raman lidar has been applied to the measurement of a number of atmospheric molecules, including H_2O, SO_2, and CO_2. The most significant limitation of the technique derives from the low cross section for Raman scattering. Thus, the application requires the use of high-power lasers, large telescopes, and long integration times and is generally limited to measurements where high molecular concentrations are present (Goldsmith et al. 1998).

Our experimental lidar system (Lay Ekuakille and Trotta 2006) uses an XeF excimer laser (lambda Physik LPX 210i), as illustrated in the photograph in Fig. 7.6.

Fig. 7.7 Lidar Newtonian telescope

An unstable cavity is applied to the laser source to obtain a low laser beam divergence. The laser, equipped with an unstable cavity, sends pulses of 150 mJ, of 30 ns duration, $30 \times 20\ \mathrm{mm}^2$, and approximately 0.3 mRad of divergence. The collected backscattered radiation is obtained by a Newtonian telescope (Fig. 7.7) whose primary mirror is 3 cm in diameter and has a focal length of 120 cm. The gathered radiation is spatially filtered by a diaphragm and separated in three different spectral channels corresponding to water vapor Raman radiation, elastically scattered radiation, and nitrogen scattered Raman radiation. A receiving optics, as shown in Fig. 7.8, plays a key role in collecting useful signals.

A second example of an experimental measurement setup is related to the measurement of exhaust using an optical system. Optical investigations of the exhaust emitted by internal combustion (i.c.) engines and a stationary burner were performed

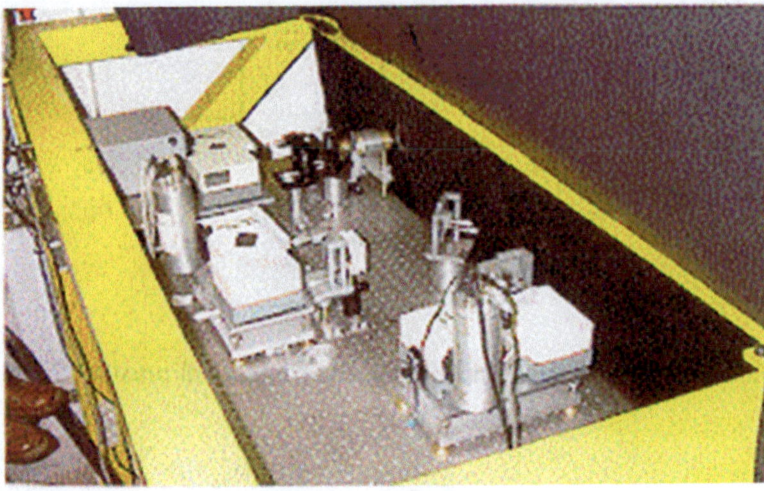

Fig. 7.8 Receiving optics

in order to assess their relative role as sources of organic matter in the atmosphere. Extinction spectra of air-diluted exhaust in the 200 ± 400 nm u.v. band revealed the expected existence of trace gases (NO, NO_2 and SO_2) and carbonaceous particulate matter (soot). In addition, after subtracting the absorption contribution from known species, a strong residual absorption band remained below 250 nm, which was attributed to organic aromatic matter, involving no more than two aromatic rings. A set of ex situ extinction and laser-induced fluorescence (LIF) experiments were carried out on condensed combustion-water samples. Extinction measurements from the water samples showed absorption spectra similar to those observed from air-diluted samples, which were attributed to low-volatility organic compounds because they were trapped in the condensed phase. Combining the indications of extinction data for both air-diluted and condensed samples, it was suggested that the absorbing species might be molecular clusters of one/two aromatic rings. LIF spectra from condensed samples evinced two fluorescence bands, centered above 300 and 400 nm, whose intensities correlated with the combustion regimes. Analogous optical analysis on rain samples collected in an urban area showed that rain absorption and fluorescence spectra are similar to those found in condensed exhaust samples, which is consistent with the prevailing contribution of i.c. engines to urban air pollution. The combined experimental data suggest that the absorbing and fluorescent species trapped in the condensed samples were organic (aromatic) compounds involving mostly one ± two aromatic rings structural units since they did not absorb above 250 nm. The overall molecular weight of the trapped material was likely heavy as they showed low volatility. In this paper, the Lambert–Beer law was used to perform the experimental setup. The Lambert–Beer law describes the correlation between the absorption behavior of a substance as well as the concentration and the layer thickness of this substance in solution (Kittelson 1997). With the help of a diode array spectrometer assembled on the optical bench, the concentration dependency of the extinction was investigated at the wavelength of the absorption maximum of a colored substance. If sample solution having a layer thickness d is irradiated with light, the initial intensity I_0 of the radiation is reduced by interaction with the substance. The reduction in intensity dI is proportional to the layer thickness (light path) d and the concentration c:

$$-\frac{dI}{I_0} = k.c.d, \tag{7.9}$$

where

I = radiation intensity
k = proportionally factor
c = concentration
d = layer thickness

Integration supplies the Lambert–Beer law:

$$I = I_0 e^{-k.c.d} \tag{7.10}$$

or

$$\ln\frac{I}{I_0} = -k.c.d. \tag{7.11}$$

For more convenient use, the decadic logarithm (Ig) is used and additionally defines extinction E as follows:

$$E = \lg\frac{I_n}{I} = \varepsilon.c.d, \tag{7.12}$$

where

E = extinction
ε = molar (decadic) extinction coefficient
c = concentration
d = layer thickness

The expression Ig I_0/I is a measure of the absorption behavior of the substance and is termed extinction. The substance-specific constant ε is a function of the wavelength and has the dimensions $\mathrm{I} \cdot \mathrm{mol}^{-1} \cdot \mathrm{cm}^{-1}$, where the concentration c is given in mol/l:

$$E_\lambda = \varepsilon_\lambda.c.d. \tag{7.13}$$

As can be seen from (7.13), as a result of the wavelength dependency of ε, extinction is also wavelength-dependent. For a constant wavelength l and a constant layer thickness *d*, there is a linear correlation between the substance concentration *c* and the extinction *E*. It also follows from (7.13) that for constant wavelength *l* and constant concentration *c* there is a linear correlation between the extinction *E* and the layer thickness *d*. Particulate matter (PM) emissions from diesel engines have been reduced by more than an order of magnitude since the inception of mandated emission regulations (Kittelson et al. 1999), and an additional order-of-magnitude reduction is required in the United States beginning in 2007. This new regulation has created a pressing need for improved instrumentation for PM measurement, particularly with regard to increased sensitivity and temporal response. Optical diagnostics are well suited for this task, where a high-energy pulsed laser can be used to obtain a variety of complementary measurements characterizing PM emissions. Researchers from the Combustion Research Facility at Sandia National Laboratories set out to innovatively combine and modify known optical diagnostic techniques to deliver a set of highly useful parameters:

- Particle volume concentration
- Particulate matter volatile fraction
- Number density and diameter of primary particles
- Geometric mean and standard deviation of the number of primary particles per aggregate
- Mass fractal dimension
- Radius of gyration of aggregated primary particles

Before describing the experimental setup carried out by us, it would be suitable to give a survey of the traditional way to measure PM by means of opacity. Nowadays, opacity is measured in a technical structure. That is a main disadvantage since it is important to get on-board data. There are usually three methods used to characterize PM excluding laboratory instrumentation that can determine PM section (Van de Hulst 1981):

Traditional method uses smoke-meter that determine opacity at exhaust according to the degree of light ray darkness;
It is possible to estimate PM concentration of PM from comparison between special paper – filter darkness with a n appropriate graduated scale;
Mass measurement of PM deposited on appropriate filter.

Opacity is defined as the ratio of transmitted light by a source that, after crossing exhausted gas, cannot reach a receiver. Thus, an opacimeter is an instrument used to measure, in a continuous way, opacity values. The laboratory where experimental tests were carried out has a FIAT 1,929 cm^3 direct-injection, turbo diesel engine. Many tests were performed in one of the following ways: constancy of revolutions, of torque, and of alpha. Figure 7.9 illustrates the main blocks that belong to the system architecture. Figure 7.10 shows the details of the measurement system.

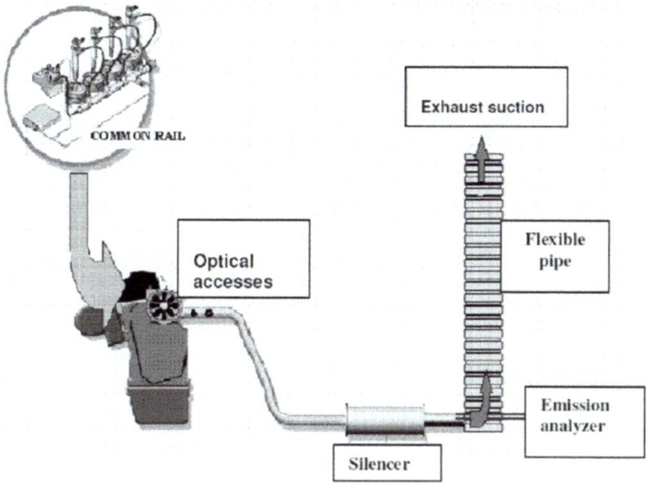

Fig. 7.9 Main circuit blocks

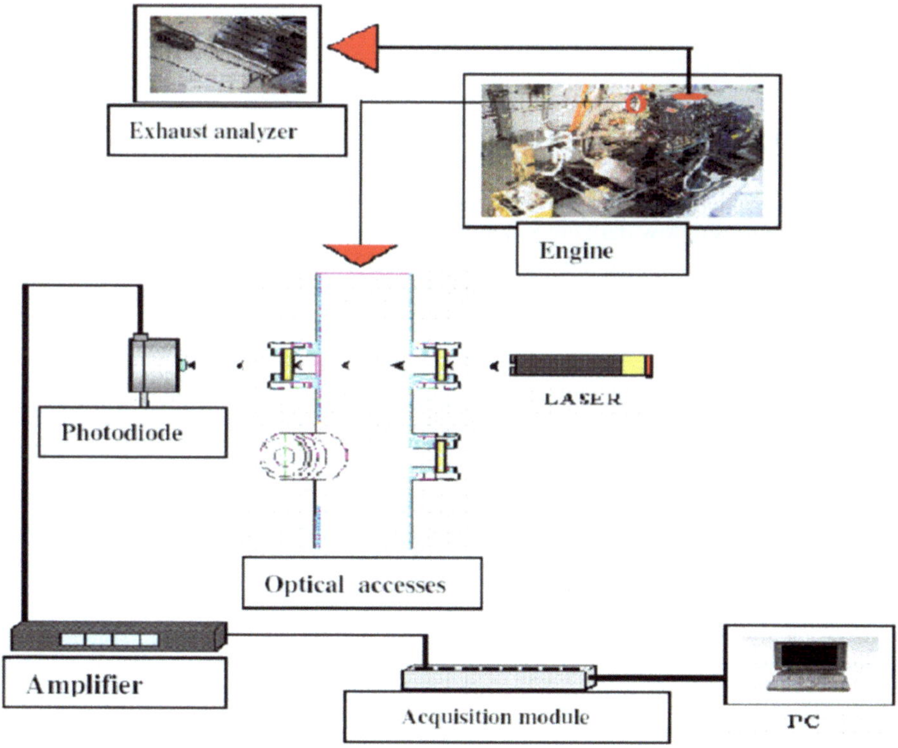

Fig. 7.10 Circuit block diagram for measurements

7.3.1 Emission Analyzer

We used an AVL Dicom 4000 analyzer that allows online characterization of five gases (CO_2, CO, NO_x, HC, O_2) and controls opacity values in diesel engines. An exhaust gas monitoring probe and an opacity measurement probe were directly inserted into a propulsion exhaust system in a certain area with constant section without pipe fitting and elbow, so that measurements could not be affected (Jones 1999).

7.3.2 Silencer and Optical Access

Extinction measurements were performed using an optical-accessed silencer with the simultaneous acquisition of opacity and concentration through the AVL Dicom 4000. The silencer was immediately assembled downhill, after the turbocompressor.

7.3.3 Photodiode

The light ray, emitted by a laser diode, after passing crosswise the optical-accessed silencer, was detected by a device able to determine its intensity. A silicon 13-DSI007 Melles Griot photodiode was used with an active area of 10 mm^2 under conditions of low dark current and high linearity of response. The photodiode was put inside an appropriate case (Universal Detector Mount 13 DMA, Melles Griot) and electrically connected to a BNC connector coming from an acquisition card channel.

7.3.4 Light Source

Extinction measurements were performed using different laser diodes with operating ranges between 635 and 785 nm. The first measurement campaign was conducted using a 55–412 Edmund Optics Synchro laser diode with maximum power of 5 mW at 635 nm; a complete characterization was executed at nontraditional wavelength at 1,523 nm, in near-IR, in order to make a deep assessment of what happens outside the established range. Hence, a 5 mW Melles Griot He-Ne 05-LPR 151 laser was used.

7.3.5 Operating Modes

The experimental stage was performed by varying some parameters related to injection and determining their influence on emissions assessed at exhaust. Experimental measurements were carried out at constant values of torque, revolution, main injection anticipation, and injection pressure. For every test we determined the photovoltage detected by the photodiode and, simultaneously, opacity values and gas concentration according to Fig. 7.9.

7.4 Digital Signal Processing Techniques and Hardware

Quantitative measurements (Klett 1981) of aerosol optical properties using a lidar system that measures only aerosol backscatter require accurate system calibration and assumptions regarding aerosol optical properties (Fernald 1984). Lidar systems that can alleviate some of these restrictions by using a multiangle integral solution of the lidar equation to solve for both backscatter and extinction. However, this method requires horizontal homogeneity of the aerosol. Since the photomultiplier tube contains a variety of noise pulses in addition to the signal pulses representing photoelectrons, simply counting the pulses without some form of noise elimination will not result in an accurate measurement. Despite the optical filters included in the experimental apparatus, there is a need for further filtering using signal digital filtering. In the common technical literature, a transfer function $H(z)$ of the form

$$H(z) = \frac{b_0 + b_1 z^{-1} + \ldots + b_n z^{-n}}{1 + a_1 z^{-1} + \ldots + a_n z^n} \tag{7.14}$$

is equivalent to the standard difference equation

$$y(k) = \sum_{i=0}^{n} b_i u(k-i) - \sum\nolimits_{i=1}^{n} a_i y(k-i). \tag{7.15}$$

There are two primary classes of digital filters. If in (7.14) and (7.15) $a_i = 0$, $i = 1,2...n$, then the filter is a finite-impulse response (FIR) filter. Otherwise it is an infinite-impulse response (IIR) filter. FIR filters have the following main properties that can be adapted to our purposes:

1. A linear phase response is easily obtained. Linear phase filters are called phase distortionless. A linear phase implies a pure time delay. These filters are useful in applications where frequency dispersion effects caused by nonlinear phase response must be minimized, such as in lidar signal acquisition and transmission.
2. Stability of the filters is guaranteed. Thus these structures are often used as adaptive filters in which the coefficients of the filter are modified in accordance

with the incoming data. In our applications, coefficients are modifiable according to acquisition weather conditions.
3. The window design method can be used as a reasonable approximation if efficiency of the design is not critical.
4. Finite register effects are inherently simpler to analyze and less consequence than in IIR filters.

Aerosol extinction is the final coefficient that represents a goodness factor for the method used in this work. The aerosol coefficient $\alpha_{\lambda_R}^{aer}$ at the Raman-shifted wavelength λ_R is given by

$$\alpha_{\lambda_R}^{aer}(z) = \frac{\frac{-d}{dz}\ln\left(P_{\lambda_R}^{korr}(z)\right)}{1 + \left(\frac{\lambda_0}{\lambda_R}\right)^k}, \qquad (7.16)$$

where $P_{\lambda_R}^{korr}$ is the detected signal corrected for a range:

$$P_{\lambda_R}^{korr}(z) = \frac{P_{\lambda_r}(z)z^2}{N_R(z)\exp\left(-\int_0^z \left[\alpha_{\lambda_0}^{mol}(\varsigma) + \alpha_{\lambda_R}^{mol}(\varsigma)\right]\right)d\varsigma}, \qquad (7.17)$$

with:

λ_0 emitted wavelength
z altitude
$P_{\lambda_R}(z)$ received Raman backscatter signal from altitude z;
$N_R(z)$ number density of Raman scatterers, here N_2 molecules;
$\alpha_{\lambda_0}^{mol}, \alpha_{\lambda_R}^{mol}$ molecular extinction at λ_0 and λ_R, respectively.

This method allows one to determine the extinction coefficient independently, with the only necessary assumption being the wavelength dependence of the aerosol extinction coefficient $\alpha_{aer}(\lambda) \sim \lambda^{-k}$.

We summarize the path followed to obtain the aerosol extinction trend:

- Signal samples are acquired from the experimental apparatus described in Fig. 7.6.
- A designed filtering system is utilized to discriminate actual and significant signals from noise, instead of producing signals from average Poissonian-calculated sample vectors.
- Aerosol extinction is computed from (7.16) in two versions: numerically and functionally. The numerical method has a step trend. Extinction calculated according to the numerical method becomes zero when the altitude is 2,500 m. That is a good value according to the specialized scientific literature.

Table 7.1 Window specifications for filtering optimization

Window	a	b	c
Rectangular	1	0	0
Hann	0.5	−0.5	0
Hamming	0.54	−0.46	0
Blackman	0.42	−0.5	0.08

In this paragraph, we consider an aerosol profile obtained using an error band. The band represents an oscillation of aerosol recovering due to the different types of noise in the lidar system. An incorrect accomplishment of data acquisition and data analysis can cause a miscalculation in the estimation of the aerosol coefficient and of the statistical error. For this reason, considerable care is necessary in handling data in order to retrieve the extinction coefficient profile coming from Raman signals (Whiteman 1999). The results reported here reflect the noctitime operations using the aforementioned XeF excimer laser. Many simulations were performed to characterize the output signal from a multichannel scaler (MCS). Digital filters were used to obtain better (the same) results than those obtained using Poisson statistics. A window design of FIR filters was implemented to remove noise from useful signals containing water vapor.

The most widely used adjustable window function is the Kaiser window function. Kaiser found that a near-optimal window function could be formed using the zeroth-order modified Bessel function of the first kind. The adopted window of FIR filter design started with the design of a least-squared-error approximation. If the desired filter has a basic-pass response, the impulse response of the optimal filter is

$$\hat{h}_d[n] = \frac{\sin(\omega_0 n)}{\pi n}. \tag{7.18}$$

the shifted and truncated version is

Relationship (7.18) is valid for $0 \leq n \leq \mathrm{L}-1$; otherwise, $h(n)$ is zero for $M = (L-1)/2$. The truncation was obtained by multiplying (7.16) by a rectangle function. Since that is what causes the Gibbs effect, we will multiply by a window function that has a smoother Fourier transform with lower side lobes. One method of smoothing the ripples caused by the sinc function is to square it. This results in the window's being a triangle function, also called a Barlett window. The four generalized cosine windows used in our experimentations are given by

$$h(n) = \begin{cases} \dfrac{\sin(\omega_0(n-M))}{\pi(n-M)} \\ 0 \end{cases}. \tag{7.19}$$

Relationship (7.19) is valid for $0 \leq n \leq \mathrm{L}-1$, otherwise 0. The names of the used windows and their parameters are illustrated in Table 7.1.

A more flexible and general window, used in this paper, is the Kaiser window, which is given by

$$W[n] = \begin{cases} \dfrac{I_0\left(\beta\sqrt{1-[2(n-M)/(L-1)]^2}\right)}{I_0(\beta)} \\ 0 \end{cases}. \tag{7.20}$$

Relationship (7.20) is valid for $0 \le n \le L-1$, otherwise zero, where $M = (L-1)/2$, $I_0(x)$ is the zeroth-order modified Bessel function of the first kind, and β is a parameter to adjust the width and shape of the window. The generalized cosine windows have no ability to adjust the tradeoff between the transition bandwidth and overshoot and therefore are not very flexible filter design tools. The Kaiser window, however, has a parameter β that does allow a tradeoff and is known to be an approximation to an optimal window. An empirical formula for β that minimizes the Gibbs overshoot is

$$\beta = \begin{cases} 0.1102(A-8.7), \\ 0.5842(A-21)^{0.4} + 0.07886(A-21), \\ 0, \end{cases} \tag{7.21}$$

where the first line of the relationship is valid for $50 < A$, the second one is valid for $21 < A < 50$, and the third one for $A < 21$,where

$$A = -20\log_{10}\delta, \tag{7.22}$$

$$\Delta = \omega_s - \omega_p, \tag{7.23}$$

$$L - 1 = \frac{A-8}{2.285\Delta}, \tag{7.24}$$

with δ being the maximum ripple in the passband and the stopband. Because the Bartlett, Hanning, and Blackman windows are zero at their endpoints, multiplication by them reduces the length of the filter by 2. To prevent this shortening, these windows are often made $L + 2$ in length. This is not necessary for the Hamming or Kaiser windows. These windows can be used on the classical ideal low-pass filter given and on any ideal response to smooth out a discontinuity.

Figures 7.11 and 7.12 illustrate recovered signals in the time domain and frequency domain, respectively. The foregoing representations do not deceive some body since the actual reasons are to remove noise by characterizing the lidar output as an alternative way by respect to Poisson average. In Fig. 7.12 different FIR windows, not only the Kaiser window, are shown to represent their impact in lidar recovering data. Although in Fig. 7.11 we illustrated just one FIR window, we did so to simplify the representation. It is clear that the NED approach is the best approach for the present lidar signal since it allows a maximum reduction of noise. That is one of the advantages of adjustable windows. On the one hand, in

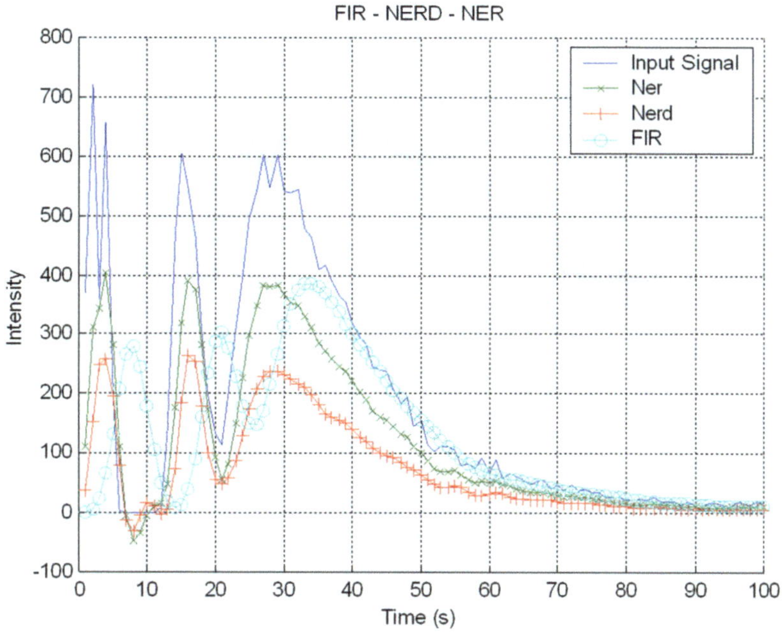

Fig. 7.11 Retrieved signals for various techniques in time domain

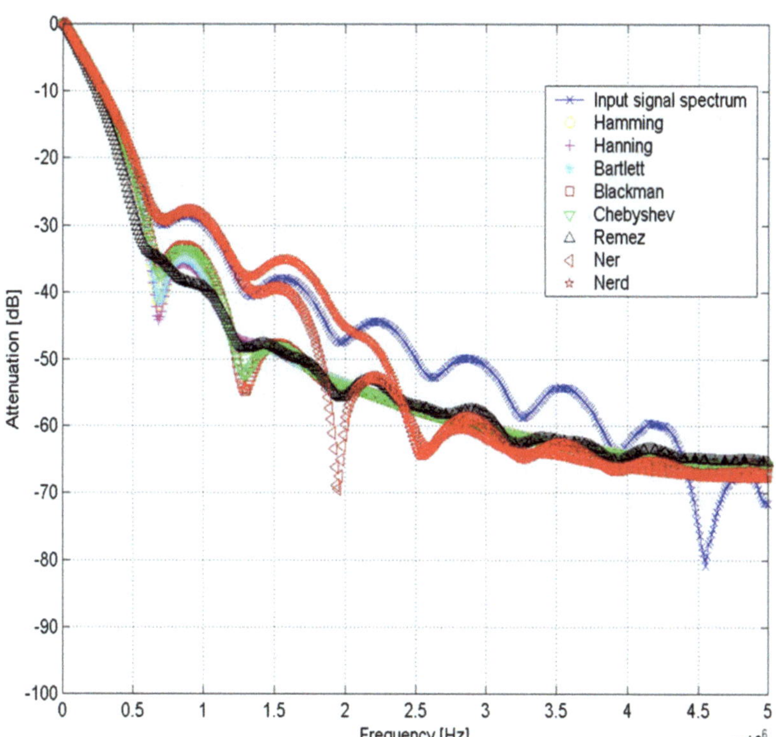

Fig. 7.12 Retrieved signals for various techniques in frequency domain

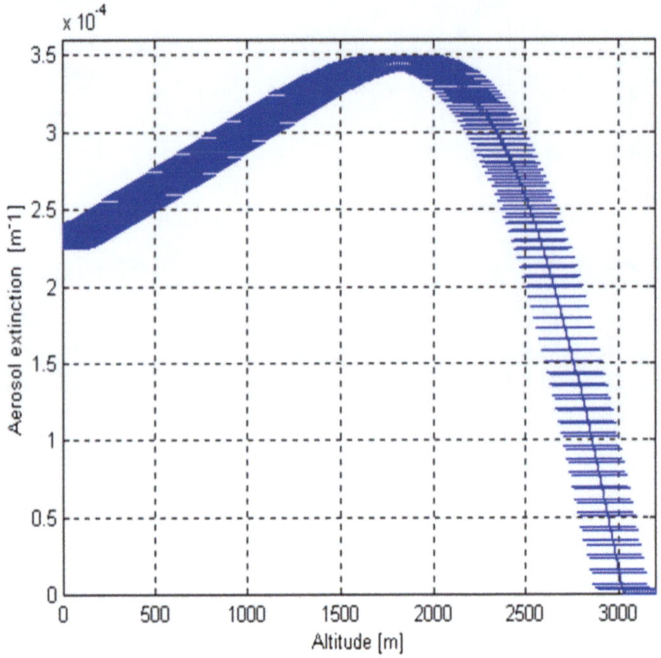

Fig. 7.13 Aerosol extinction recovery with uncertainty band

the NER case, the resulting and designed filter has a frequency response characteristic that is very nearly equal ripple about unity in the passband, $0 < f < \beta - \delta/2$, and about zero in the stopband, $\beta + \delta/2 < f < 1.0$. The gain at zero frequency is slightly different than unity, but within $\pm\,\varepsilon$ of unity. We noticed that the reduction in the passband ripple and the increase in the stopband attenuation were achieved at the expense of a widening transition band. To narrow the transition band, a larger number of terms was used; on the other hand, the NERD derivative filter design is characterized by approximately equal passband and stopband magnitude errors. The resulting designs are very nearly the most efficient possible while remaining consistent with the filter performance specifications. The error of the extinction coefficient (Hamming 1983) is mainly determined by the statistical error of the Raman return signal. This error is assumed to be the square root of the absolute count rate and is calculated via the law of error propagation in the extinction coefficient. A rapid change in the aerosol extinction coefficient during the averaged period induces a significant error in the mean extinction coefficient. To avoid such errors, only periods with approximately constant aerosol backscatter signals were averaged. Aerosol extinction (Fig. 7.13) is the final coefficient of this work that represents a merit factor for the method used in this paper to characterize

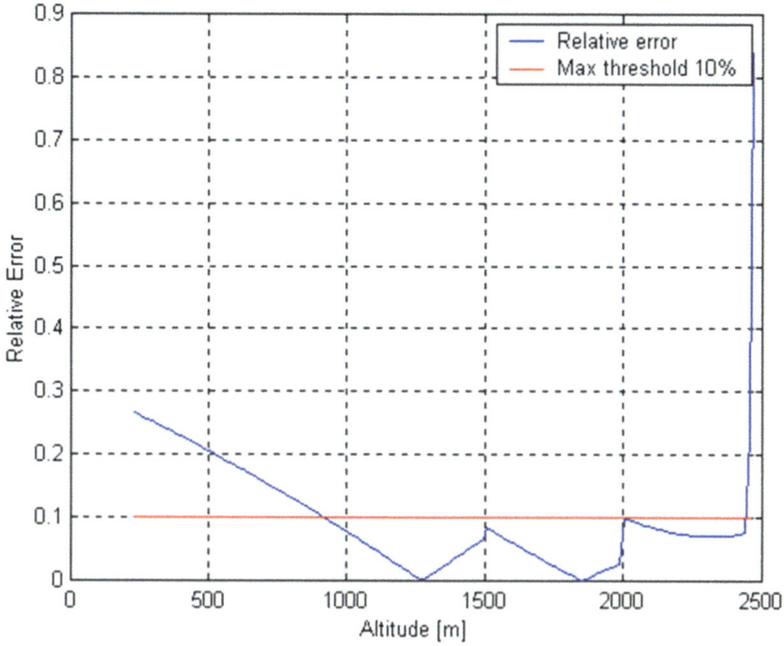

Fig. 7.14 Error determination

the sensor. The band of uncertainty, plotted in Fig. 7.13, was computed from absolute uncertainty determined from (7.23):

$$U_{aer} = \left|\frac{\partial \alpha_{aer}}{\partial (z^2 s)}\right| U_{z^2 s} + \left|\frac{\partial \alpha_{aer}}{\partial \rho}\right| U_{\rho} + \left|\frac{\partial \alpha_{aer}}{\partial \alpha_{ray}}\right| U_{ray}, \tag{7.25}$$

where U represents the absolute uncertainty of the extinction coefficient. Practically, we added relative errors of z^2s [numerator of under derivation of (7.16)] and of ρ (denominator); once we passed to absolute values, we added the absolute uncertainty of $\alpha^{ray}{}_{\lambda_L}$. By analyzing numerical values, we noticed that the contribution provided to the overall error of $\alpha^{aer}{}_{\lambda_L}$ from term present in derivation operation has different orders less than the contribution of molecular extinction. That justifies the achievement of a constant amplitude band with respect to z; that is, the error of s is the only one considered variable with respect to that parameter.

To validate the results produced by the present research, we calculated the relative error plotted in Fig. 7.14 with a Matlab routine by filtering, layer by layer along altitude z, (7.16). The error in the derived aerosol extinction at 351 nm using a Raman nitrogen signal is more or less 10 % if k, as reported in (7.16), varies between 0 and 2 when an assumed value of $k = 1$ is used. This error reduces to 6 % when a Raman oxygen signal is used because of the smaller difference between the

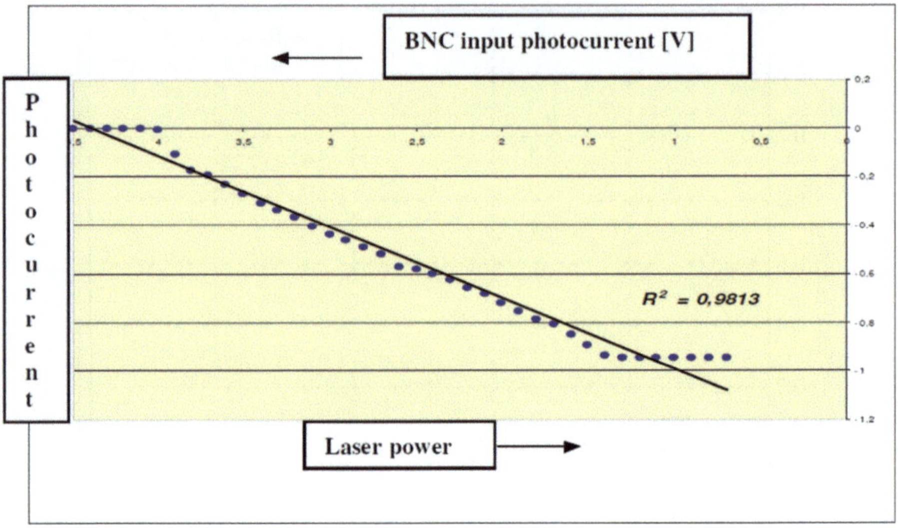

Fig. 7.15 Photocurrent [mA] versus detected light power

outgoing laser wavelength and the return Raman wavelength. According to the scientific literature the relative error must be approximately 10 % near 2,500 m of altitude, as shown in Fig. 7.14.

For exhaust monitoring, data processing was very simple since before collecting data, the calibration stage was performed; in particular, it was necessary to verify that the silicon diode could give a photocurrent response that was linear and proportional to the light intensity that reached it. In other words, it was necessary to avoid (Reciprocating 1994) that the silicon photodiode works in saturation zone, and the laser power used must be suitable for this purpose. The photodiode response curve is linear in the range of the light power used during the experimental stage. A trend of the photocurrent versus applied voltage is illustrated in Fig. 7.15. Each plotted value of the photovoltage was obtained from 25,000 samples collected using Labview software for every engine regime reached; Fig. 7.16 summarizes, in a synthetic manner, the comparison between the opacity measured via laser and that acquired through the traditional method at different laser wavelengths. The first series of extinction measurement was carried out using a 55–412 Edmund Optics Synchro laser diode at 635 nm (5 mW), varying input voltage to 2.5 V, which allowed us to obtain an emission power of 60 % with respect to the maximum power of 3 mW, while the opacimeter allowed us to obtain opacity values directly expressed in percentages, and the average values of the photovoltage obtained for each engine regime were postprocessed using Lambert– Beer'law and transformed in percentages of opacity values. In this way we made an immediate comparison. All of this was repeated for all laser wavelengths until the maximum one.

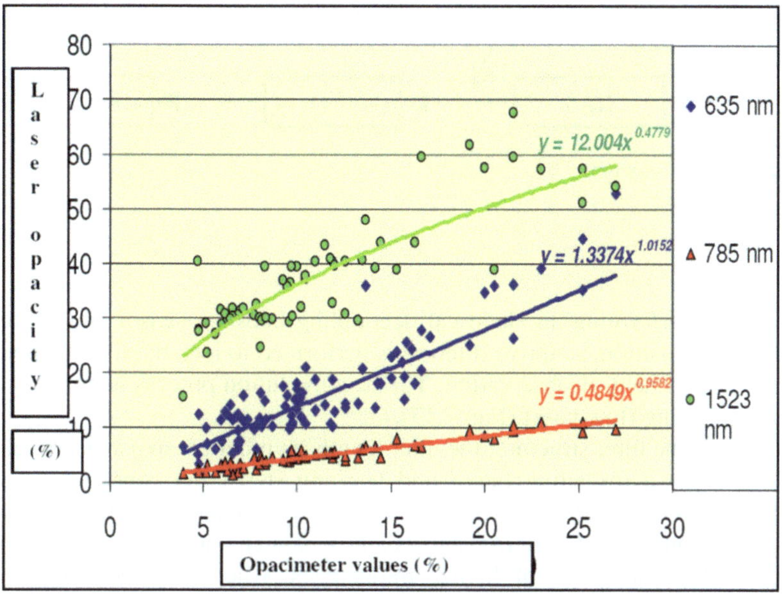

Fig. 7.16 Resulting correlations for three laser wavelengths and their trends

7.5 Advanced Techniques

A lidar observation may be considered a 1 × N image, by the Richardson-Lucy algorithm, if submitted to a deconvolution process. The measured signal *Pm*(*t*) is taken as the initial guess $P^{TM}(1)$ for the iteration, and negative values due to noise must be set to zero to guarantee its convergence (Harsdorf and Reuter 2000). In a complex environment at lidar signal is superposed noise *N*(*t*):

$$P_m(t) = P(t) + N(t) = (R(t) \times P_\delta(t)) + N(t), \tag{7.26}$$

where *P*(*t*) is a lidar observation after the deconvolution process and is strongly dependent on the geometrical and optical characteristics of the sensor, and $P_\delta(t)$ is the ideal observation when the response function of the lidar sensor is a Diràc distribution, like an ideal laser pulse. From (7.26), the additive noise term *N*(*t*) makes direct deconvolution impossible. The aim of the algorithm is to evaluate the performance of the system, as is done in what follows, where DAS is the system that performs the data acquisition, which in any case may be considered "all-in-one" with the lidar bulk. DAS will guarantee a "nondemolition" extraction of

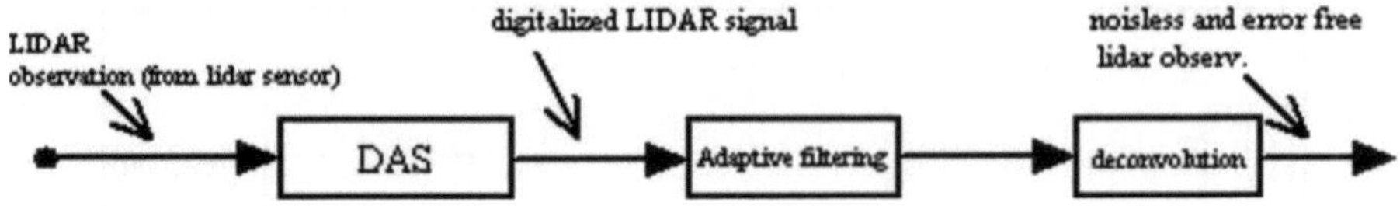

Fig. 7.17 System structure

information, performing an electrical decoupling between sensor and processing unit. After acquisition, adaptive filtering is performed to lock the filter response on the noisy component of observation. The deconvolution process is now error free due to prefiltering (low pass) (Fig. 7.17).

We design the filter structure: the filter length (number of taps) and various filter parameters like a forgetting factor and learning step (gain constant). Then the system performance will be investigated and specific results outlined. In general, a deconvolution process with $R(t)$ shows a low-pass characteristic, and this operation intensifies the range of higher frequencies, where $N(t)$ contributions are relevant. Consequently, a pre-low-pass filtering of observed data can be very useful in some cases. The aim is to propose a new advanced filtering scheme with an adaptive noise canceller (ANC), which, with a knowledge of a priori noise statistics and the lidar profile, optimizes a set of digital filter coefficients to adapt its impulse response to improve the SNR (Neveux et al. 2000). The choice of adaptive algorithm in a normalized least mean square (N-LMS) that updates the filter weight looking up to input signal power, making a fine tuning of impulse response.

The Matlab system simulation will now be presented. The first step is to construct a typical output lidar profile, as in Fig. 7.18. Then we superpose a zero mean Gaussian white noise $\sigma^2 = 0.08$, obtaining the results shown in Fig. 7.19. After low-pass filtering with ANC (Fig. 7.20) and using a deconvolution process by the Richardson–Lucy algorithm that yields to Fig. 7.21. The ANC system performs low-pass filtering on a noisy lidar signal and its effects can be seen by the filter frequency response, which shows a low-pass characteristic according to Fig. 7.22. These results are obtained using an ANC system with 16 taps and a gain factor (μ) of 0.01.

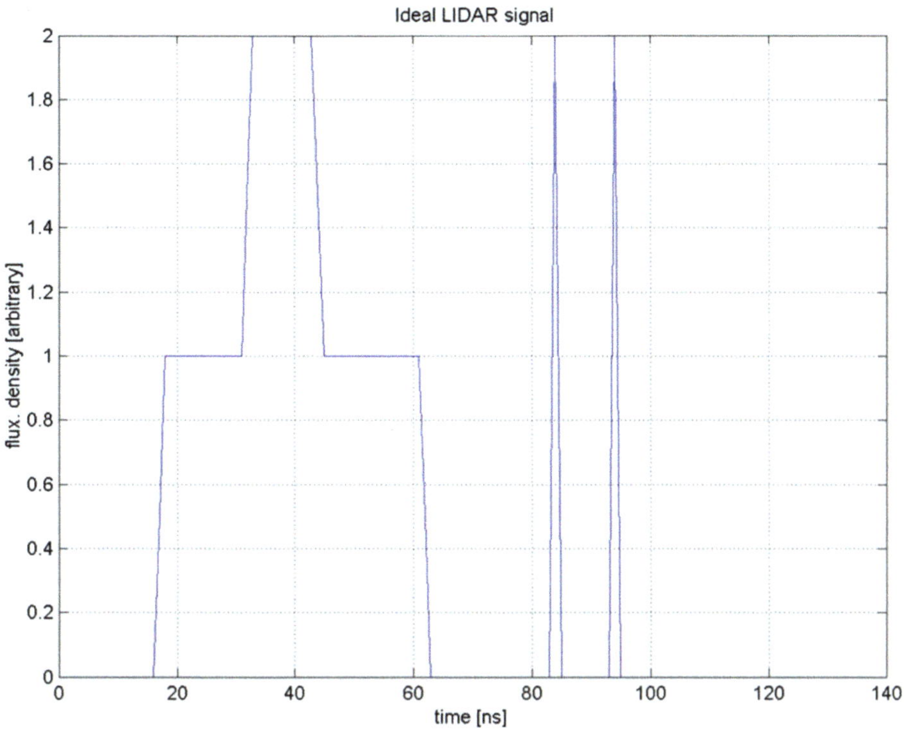

Fig. 7.18 Typical ideal lidar profile

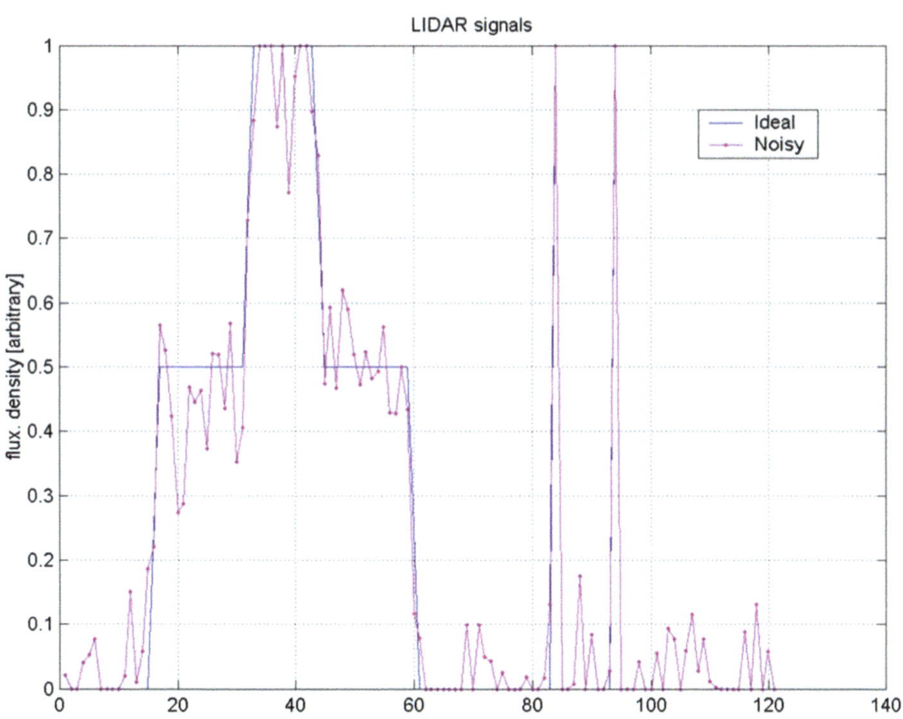

Fig. 7.19 Comparison: ideal and noisy lidar signal

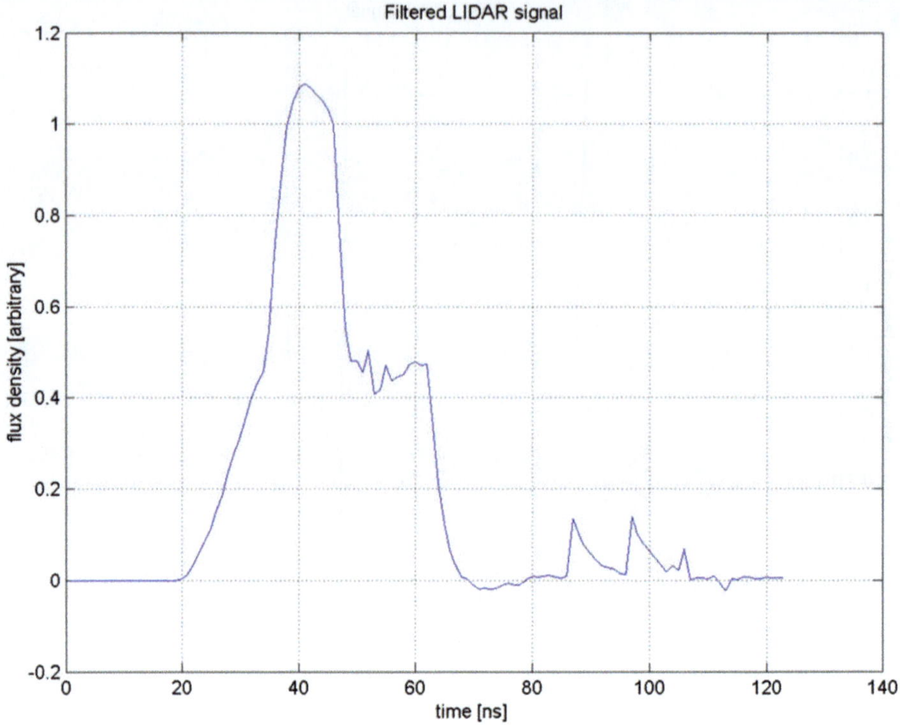

Fig. 7.20 Lidar signal filtered by ANC

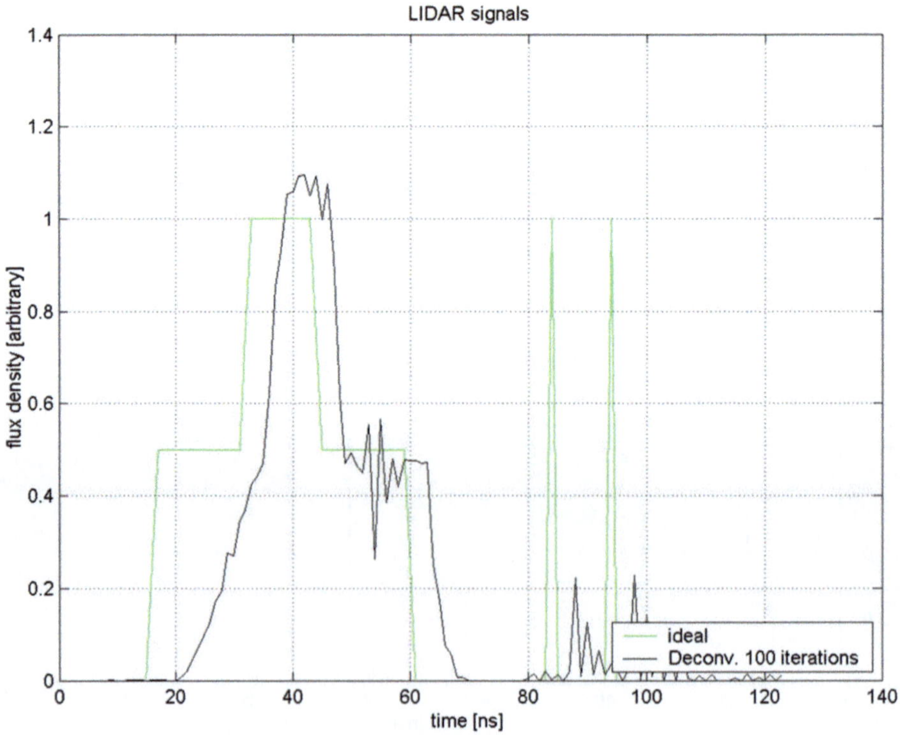

Fig. 7.21 Comparison between ideal and reconstructed lidar signals

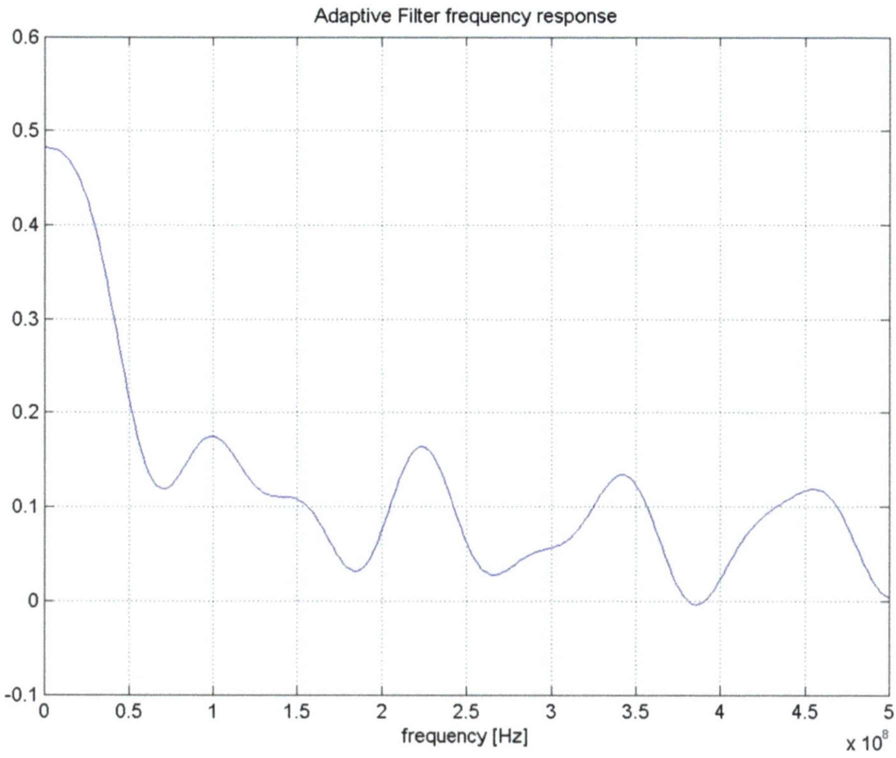

Fig. 7.22 Low-pass characteristic of adaptive filter

Acknowledgments The author gratefully acknowledges the technical support of P. Carlucci, A. Pascali, D. Laforgia, A. Ficarella, M.R. Perrone, and F. De Tomaso.

References

Corney, A.: Atomic and Laser Spectroscopy. Oxford University Press, London (1977)

Dally, J.W., Riley, W.F., McConnell, K.G.: Instrumentation for Engineering Measurements, 2nd edn. Wiley, New York (1993)

Fernald, F.G.: Analysis of atmospheric lidar observations: some components. Appl. Opt. **23**, 625–653 (1984)

Goldsmith, J.E.M., Blair, F.H., Bisson, S.E., Turner, D.D.: Turn-key Raman lidar for profiling atmospheric water vapor, clouds, and aerosols. Appl. Opt. **37**(21), 4979–4990 (1998)

Griffiths, P.R.: Chemical Infrared Fourier Transform Spectroscopy. Wiley, New York (1975)

Hamming, R.W.: Digital Filters, 2nd edn. Prentice-Hall, Englewood Cliffs (1983)

Harsdorf, S., Reuter, R.: Stable deconvolution of noisy lidar signals. In: Proceedings of EARSeL-SIG-Workshop LIDAR, FRG, Dresden, 16–17 June 2000

Hu, M., et al.: Dark-field microscopy studies of single metal nanoparticles: understanding the factors that influence the linewidth of the localized surface plasmon resonance. J. Mater. Chem. **18**, 1949–1960 (2008)

Jones, A.R.: Light scattering for particle characterization. Prog. Energy Combust. Sci. **22**, 1–53 (1999)

Kittelson, D.B.: Engines and nanoparticles: a review. Aerosol Sci. **29**, 575–588 (1997)

Kittelson, D.B., Abdul-Khalek, I.: Formation of nanoparticles during exhaust dilution. In: EFI Members Conference: Fuel, Lubricants Engines, & Emissions, University of Minnesota, 18–20 Jan 1999

Klett, J.D.: Stable analytical inversion solution for processing lidar returns. Appl. Opt. **20**, 211–220 (1981)

Lay Ekuakille, A., Trotta, A.: Aerosol extinction uncertainty determination for a laser-based detecting system. Trans. Syst. Signals Devices **1**, 321–338 (2006)

Lee, C.K., Wu, T.W.: Differential laser tnterferometer for nanometer displacement measurements. AIAA J. **33**(9), 1675–1680 (1995)

Lohmann, P., Koch, A., Schaeffer, M.: Approaches to the Filtering of Laser Scanner Data, vol. 33, part B3, pp. 540–547. International Archives of Photogrammetry and Remote Sensing, Amsterdam (2000)

Nawrocki, W.: Measurement Systems and Sensors. Artech House, Norwood (2005)

Neveux, P., Sekko, E., Thomas, G.: A constrained iterative deconvolution technique with an optimal filtering: application to a hydrocarbon concentration sensor. IEEE Trans. Instrum. Meas. **49**(4), 852–856 (2000)

Northrop, R.B.: Introduction to Instrumentation and Measurements. CRC, Boca Raton (2005)

Okada, K., Aoyagi, M., Iwata, S.: Accurate evaluation of Einstein's A and B coefficients of rovibrational transitions of carbon monoxide: spectral simulation of $\Delta v = 2$ rovibrational transitions in the solar atmosphere observed by a satellite. J. Quant. Spectrosc. Radiat. Transf. **72**(6), 813–825 (2002)

Pelkim Elmer.: FT-IR, Spectrum 2000 User's Reference, UK (1995)

Reciprocating Internal Combustion Engines -Exhaust Emission Measurement- Part 3, SAE Standard ISO8178-3 (1994)

Van de Hulst, H.C.: Light Scattering by Small Particles. Dover Publications, New York (1981)

Whiteman, D.: Application of statistical methods to the determination of slope in lidar data. Appl. Opt. **38**(15), 3360–3369 (1999)

Chapter 8
Noise

Abstract Noise, as an undesired signal that often overlaps with the desired information contained in a signal, must be removed or at least reduced to an acceptable level. Before designing a filtering system it is important to know the source of noise and its quality. Its representation using electrical circuits is a matter of interest in many fields of applied engineering. It can be represented as a one-dimensional and a bidimensional signal. The latter case is interesting in the field of image processing. The measurement of noise is not easy even when we can use a specific modeling.

8.1 Definitions and Sources

Optical waves, which are used in engineering applications, contain noisy components; in particular, lasers display noise in amplitude and frequency. Many measurements performed using lasers are affected by noise mostly in intensity with respect to frequency. Noise is an undesired component that is included in the signal containing information to be preserved, but it acts as a disturbing phenomenon. Intrinsically speaking, noise happen mostly due to the following reasons (Haus et al. 1960):

- Thermal agitation process and Poissonian process (bruit de grenaille) that cause white noise
- Flicker noise or pink noise that produces 1/f noise
- Brownian motion that produces random noise ($1/f^2$)
- Aging effects that allow the development of drift

The waveforms of different kinds of noise are displayed in Figs. 8.1, 8.2, 8.3, and 8.4. In each figure, the noisy components are shown according to their nature. Further definitions of noise are described in many books on physics and electronics; here we focus our attention on light, in particular on lasers.

A. Lay-Ekuakille, *Optical Waveguiding and Applied Photonics*, Lecture Notes in Nanoscale Science and Technology 10, DOI 10.1007/978-1-4614-5959-0_8,

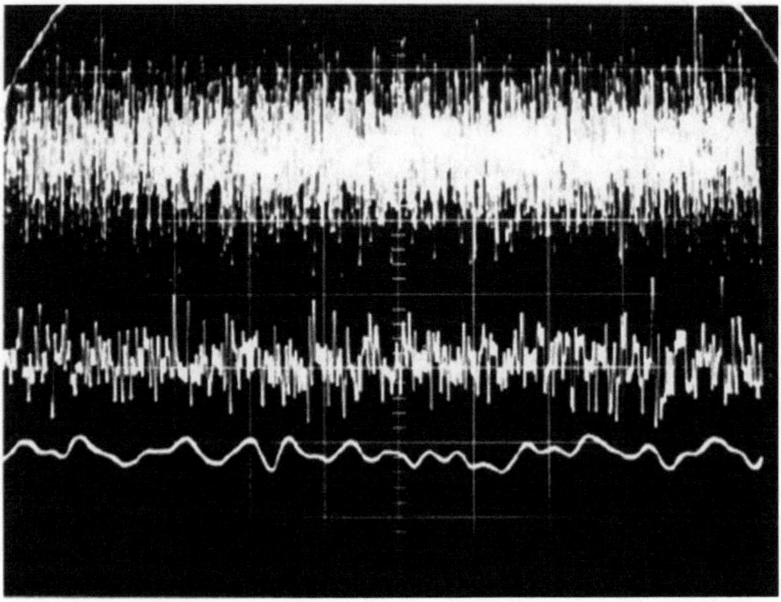

Fig. 8.1 *White* noise is shown as it appears on a scope display with *horizontal* sweep of 1 ms/cm. *Top* waveform bandwidth is dc to 200 KHz; *center* waveform, bandwidth is dc to 20 kHz; *bottom* waveform, bandwidth is dc to 2 kHz. Note that the bandwidth reduction affects both peak amplitude and the rms value (Van der Ziel 1970)

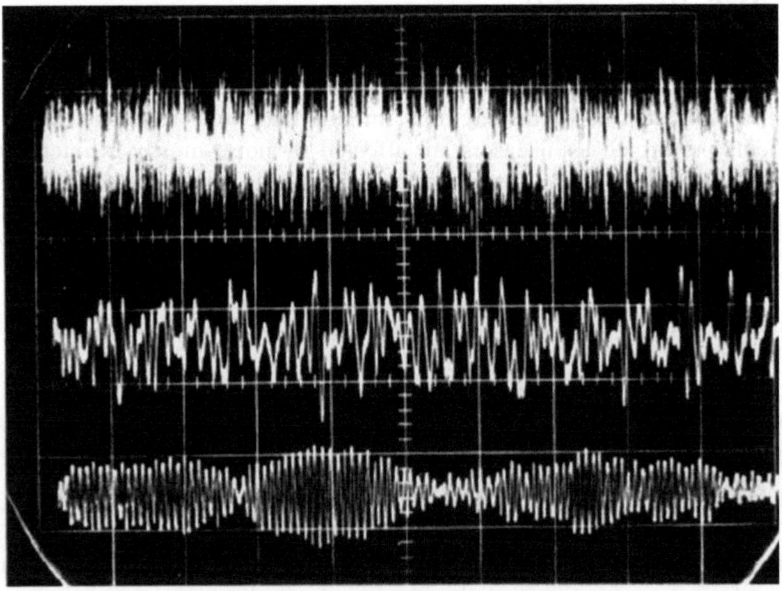

Fig. 8.2 Broadband *white* noise is shown in the *top* trace. When passed through a 2 kHz crystal filter with $\Delta f = 50$ Hz, we obtain the *middle* trace. In the *bottom* trace, the same noise is passed through a 2 kHz *RLC* high-*Q* filter with $\Delta f = 50$ Hz. Note the "ringing" effect in the bottom trace. *Horizontal* sensitivity is 5 ms/cm (Van der Ziel 1970)

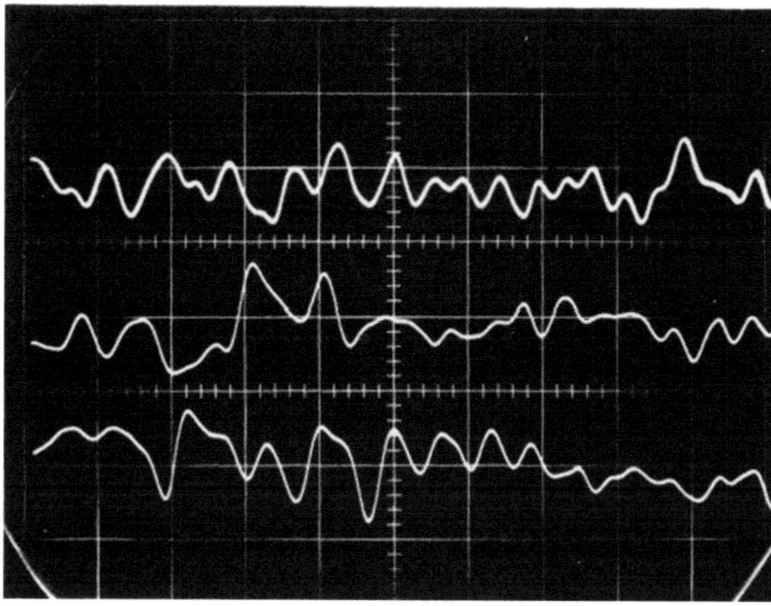

Fig. 8.3 The power content of 1/f noise in each decade of frequency is equal, as shown in these traces. *Top* waveform $\Delta f = 20$ Hz, *horizontal* 100 ms/cm; *center* waveform $\Delta f = 200$ Hz, *horizontal* 10 ms/cm: *bottom* waveform, $\Delta f = 2$ kHz, *horizontal* 1 ms/cm (Van der Ziel 1970)

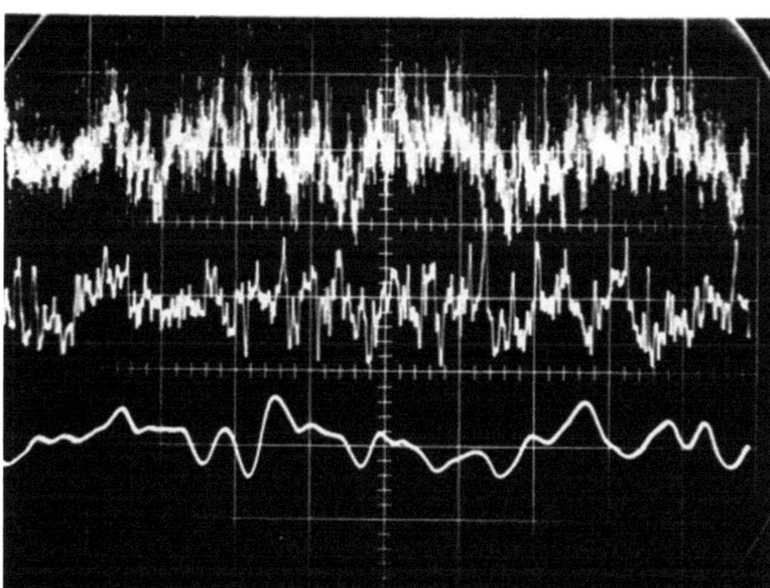

Fig. 8.4 The traces show the effect of bandwidth limiting of 1/f noise. *Top* waveform, bandwidth is 2 kHz; *middle* waveform, bandwidth is 200 Hz; *bottom* waveform, bandwidth is 20 Hz. *Horizontal* sensitivity is 50 ms/cm. Note that the peak amplitude is not proportionately reduced by bandwidth limiting (Van der Ziel 1970)

Most of these noisy signals reflect the influence of bandwidth, which is important in noise representation, especially in high-frequency applications.

8.2 Noise in Bipoles and Double Bipoles

Noise representation and modeling is crucial for understanding the sources and mechanisms of noise. Bipoles and double bipoles are important because they can be used for noise modeling. While bipoles are simple, for example resistors, if we want to model noise issues in networks, we will use double bipoles or quadripoles. One of the most interesting quantities to determine is the spectral power density beyond others such as noise factor, noise equivalent temperature, and others.

8.2.1 *Electrical Bipoles*

Let us consider a thermal noise the spectral power density (SPD) available is

$$S_n(f) = kT, \tag{8.1}$$

where k is a constant and T is the temperature. Equation (8.1) is valid for $h\nu/kT << 1$ in which ν is the radiation frequency (Fig. 8.5).

Equation (8.1) reveals the fact that spontaneous fluctuations at the terminal ends of a conductor that is kept at thermal equilibrium at temperature T are independent of conduction mechanisms, material nature, geometry, and conductor resistance R. Hence the SPD depends only upon temperature; it does not depend upon conductor resistance. The SPD (Martin 2001) is the maximum power that a source having an

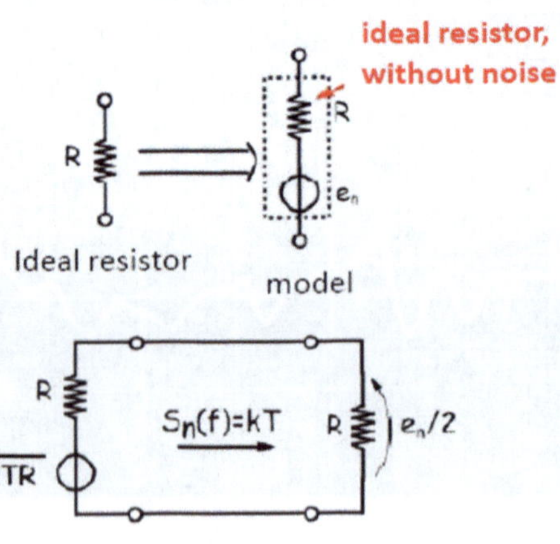

Fig. 8.5 Equivalent circuit of thermal noise

internal resistance R is able to supply to a resistive load of the same value. The SPD can also be expressed as a function of noise voltage as

$$S_n(f) = kT = \frac{e_n^2}{4R} \qquad in\ W/Hz, \tag{8.2}$$

in which

$$e^2_{R_{equ}=}4k(T_1R_1 + T_2R_2). \tag{8.3}$$

Notice that it is interesting to know the sum of uncorrelated noise contributions when resistors are in series and parallel and when they are at the same and different temperatures:

Resistors R_1 and R_2 in series

$$R_{equ} = R_1 + R_2 \;\Rightarrow\; e^2_{R_{equ}} = 4kT_0R_{equ} \tag{8.4}$$

For $T_1 = T_2$ and for $T_1 \neq T_2$ we have the following:

$$e^2_{R_{equ}} = 4k(T_1R_1 + T_2R_2). \tag{8.5}$$

Resistors R_1 and R_2 in parallel

$$R_{equ} = R_1 \| R_2 \;\Rightarrow\; e^2_{R_{equ}} = 4kT_0R_{equ} \tag{8.6}$$

For $T_1 = T_2$ and for $T_1 \neq T_2$ we have a complex formula as follows:

$$\begin{aligned} e^2_{R_{equ}} &= 4k\left[T_1R_1 \cdot \left(\frac{R_2}{R_1 + R_2}\right)^2 + T_2R_2\left(\frac{R_1}{R_1 + R_2}\right)^2\right] \\ &= 4k\frac{R_1R_2}{(R_1 + R_2)^2}(T_1R_2 + T_2R_1). \end{aligned} \tag{8.7}$$

8.2.2 Electrical Double Bipoles

Different electrical and electronic devices behave as double bipoles (e.g., amplifiers, filter banks). Once the internal structure of a double bipole is known, it is easy

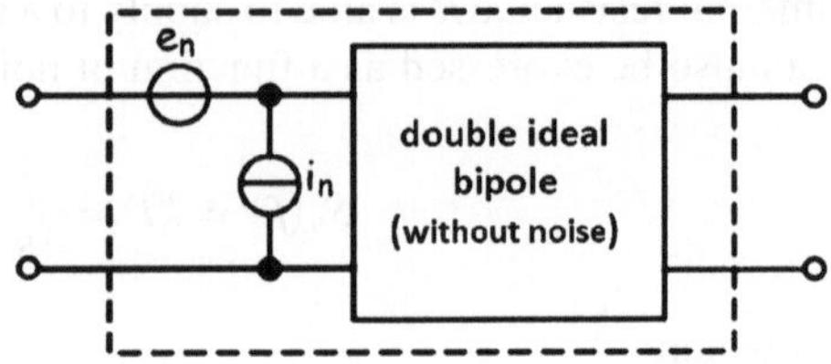

Fig. 8.6 Double real and noisy bipole

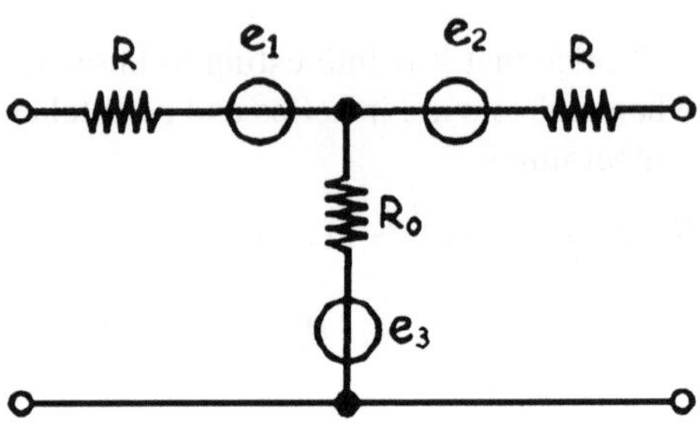

Fig. 8.7 Double bipole example

to find noise sources and describe the behavior according to noise approach. When dealing with a double bipole (Fig. 8.6), the following steps are necessary:

- Extract all internal noise contributions.
- Assign to the appropriate set of noise sources (external to the double bipole connected to the set) a level of noise such that in a global approach the system behaves as a real double bipole.
- It is usually worthwhile to put all noise-equivalent contributions to the input of the double bipole.
- For a linear double bipole, two equations are necessary to describe the relationship between input electrical quantities and output ones.

A double bipole equivalent representation includes a device (or electrical network) without an internal source of noise, a voltage generator of equivalent noise e_n, and a current generator i_n connected to the input. It is also necessary to specify the correlation between the aforementioned generators.

If we want to outline how to find the equivalent circuit for the attenuator in Fig. 8.7 in order to calculate a voltage generator of equivalent noise e_n and a current generator i_n, we need to perform the following steps:

- The value of generator e_n can be determined by assessing the open-circuit voltage v_0 at the output of both circuits (attenuator and generic model of double bipole) when the input is in a short-circuit configuration.
- The value of generator i_n can be determined by evaluating the open-circuit voltage (at the right terminal ends) of both circuits (attenuator and generic model of double bipole) when the input is in an open-circuit configuration.

The SPD can be calculated as follows:

$$e_n = e_1 + e_3\frac{R}{R_0} + e_2\left(1+\frac{R}{R_0}\right) \Rightarrow S_{e_n} = S_{e_1} + S_{e_3}\frac{R^2}{R_0^2} + S_{e_2}\left(1+\frac{R}{R_0}\right)^2$$
$$= 4kT_0R\left[\left(1+\frac{R}{R_0}\right)\left(2+\frac{R}{R_0}\right)\right], \tag{8.8}$$

$$i_n = \frac{e_3+e_2}{R_0} \qquad \Rightarrow S_{i_n} = \frac{S_{e_2}+S_{e_3}}{R_0{}^2} = 4kT_0\frac{R}{R_0}\frac{R+R_0}{RR_0}, \tag{8.9}$$

while the cross-correlation spectral density can be calculated by taking into account that e_n and i_n exhibit both contributions of noise generated by e_3 and e_2; hence, they are correlated. We can calculate the cross-spectral density reading directly from the cross-correlation function

$$S_{e_n i_n} = \frac{R}{R_0^2}S_{e_3} + \frac{R+R_0}{R_0}\frac{1}{R_0}S_{e_2} = 4kT_0\frac{R}{R_0}\left(2+\frac{R}{R_0}\right). \tag{8.10}$$

It is important to recall that the degree of correlation is often small and its contribution is neglected, and the generators are considered equivalent.

8.3 Design of Noise Removal: Measurements and Architectures

In optics noise can be removed or reduced by working on light circuitry or trying to do so after having converted light into an electrical signal. We present an example of designing a filtering system according to the second alternative. Suppose we have a laser source that is to be used for a particular purpose. The information contained in the laser beam if affected by noise must be processed by means of a complex set of filters. So we must use a photomultiplier tube (PMT), which is a device for converting a light beam into an electrical signal, that is, into an electrical current (Figs. 8.8 and 8.9).

A PMT consists of a light-sensitive cathode that emits electrons in proportion to the photons striking it. These electrons are then accelerated to the next stage, where they impinge. Each electron produces the emission of 3–6 secondary electrons. The process continues through 6–14 stages (called "dynodes"), depending on the tube type. Overall gains of one million or more are commonly attained (Keithly; http://www.hamamatsu.com) (Fig. 8.10).

A laser (http://www.ecvv.com/product/3197740.html) is used in the experimental layout described in Fig. 8.8. We start the description of the main functional blocks used in the project according to the experimental layout. A current–voltage converter

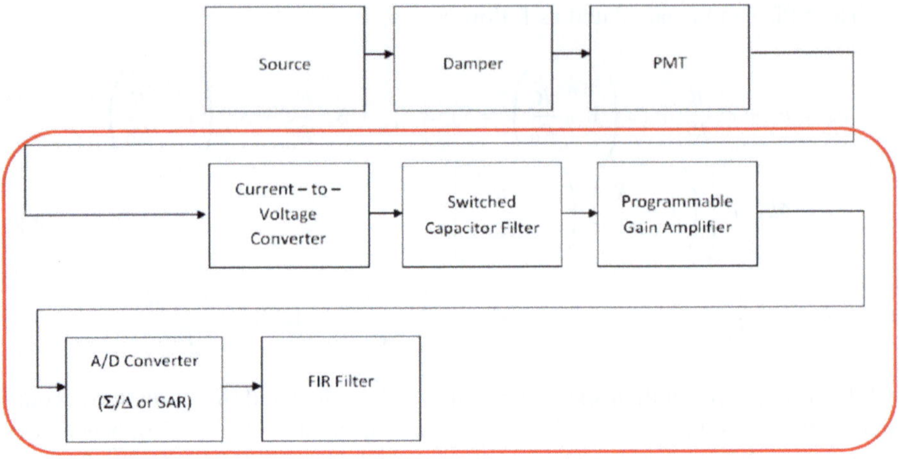

Fig. 8.8 Processing laser source for noise removal

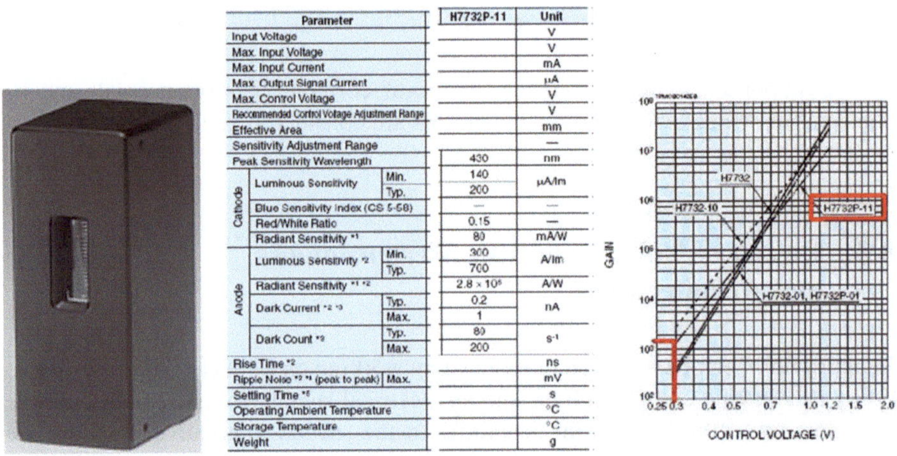

Parameter			H7732P-11	Unit
Input Voltage				V
Max. Input Voltage				V
Max. Input Current				mA
Max. Output Signal Current				μA
Max. Control Voltage				V
Recommended Control Voltage Adjustment Range				V
Effective Area				mm
Sensitivity Adjustment Range				—
Peak Sensitivity Wavelength			430	nm
Cathode	Luminous Sensitivity	Min.	140	μA/lm
		Typ.	200	
	Blue Sensitivity Index (CS 5-58)		—	—
	Red/White Ratio		0.15	—
	Radiant Sensitivity *1		80	mA/W
Anode	Luminous Sensitivity *2	Min.	300	A/lm
		Typ.	700	
	Radiant Sensitivity *1 *2		2.8 × 10⁵	A/W
	Dark Current *2 *3	Typ.	0.2	nA
		Max.	1	
	Dark Count *3	Typ.	80	s^{-1}
		Max.	200	
Rise Time *2				ns
Ripple Noise *2 *4 (peak to peak)		Max.		mV
Settling Time *5				s
Operating Ambient Temperature				°C
Storage Temperature				°C
Weight				g

Fig. 8.9 Photomultiplier tube features: used example (*left*), data sheet (*center*), and gain versus control voltage (*right*)

is designed, as indicated in Fig. 8.11, with the purpose of having an out voltage proportional to the input current. Equation (8.11) describes this behavior where I_{cc} is intended as a short-circuit current:

$$V_{out} = V_1 = -R_1 I_{cc}. \tag{8.11}$$

The second block is a switched capacitor filter since it is difficult to model this device in used version for Multisim software (http://www.ni.com/multisim/), an equivalent circuit is used that is an elliptic filter, as illustrated in Fig. 8.12.

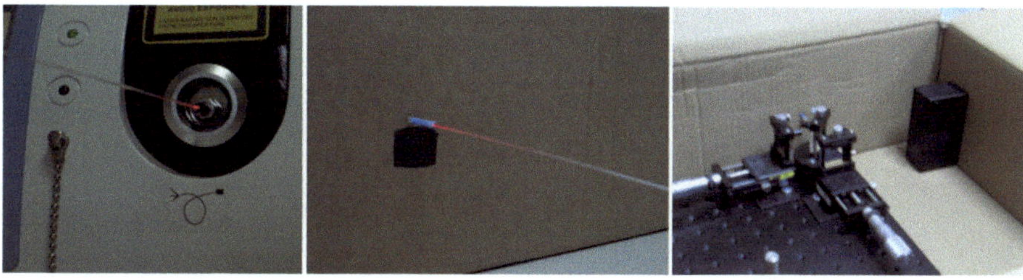

Fig. 8.10 Laser source (*left*) pointing by means of optical fiber (*center*) to photomultiplier tube (*right*)

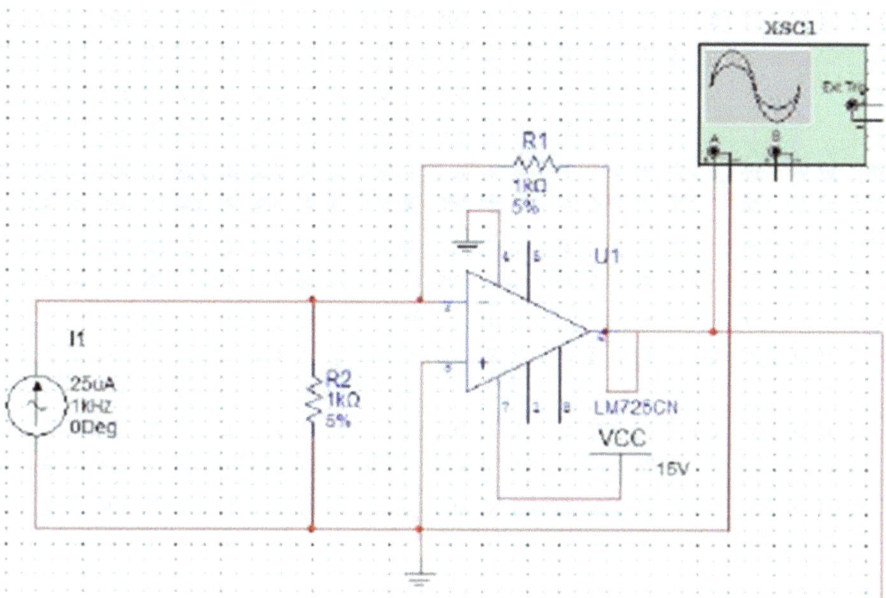

Fig. 8.11 Current–voltage converter

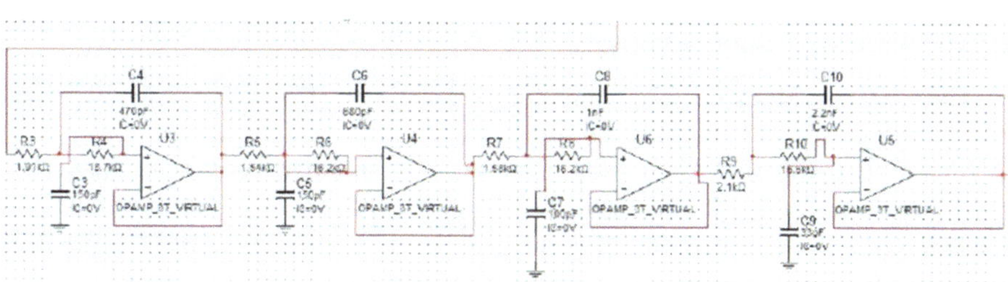

Fig. 8.12 Elliptic filter in lieu of switched capacitor filter

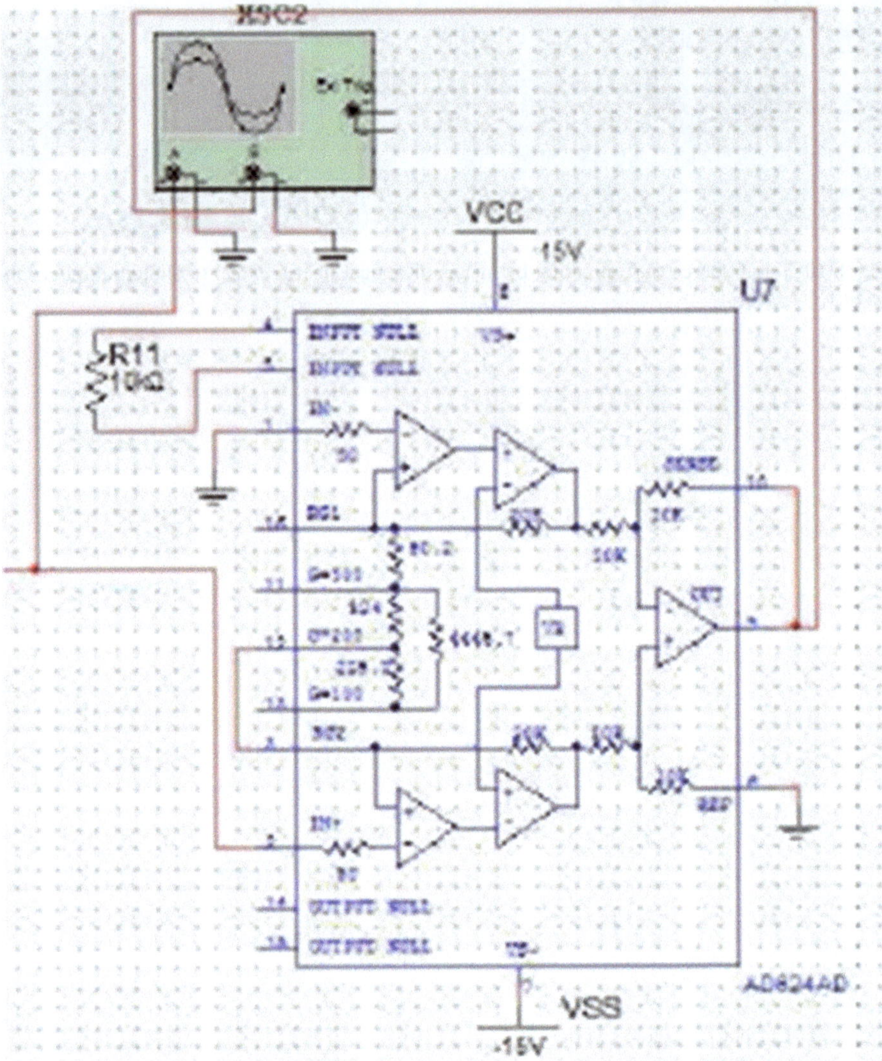

Fig. 8.13 Programmable gain amplifier circuit

The next main block is a programmable gain amplifier (PGA) (Fig. 8.13). It is used after filtering in order to amplify the signal. It provides interesting advantages since it is similar to a differential amplifier. Thanks to its internal resistors, it allows for predefined gains of 1, 100, 200, and 500. To obtain other gain values different from the previous ones, it is necessary to connect, in an appropriate way, pins by means of series/parallel combinations of resistors. With LabVIEW software (http://www.ni.com/labview/) the electrical signal after a first conversion from light, by means of

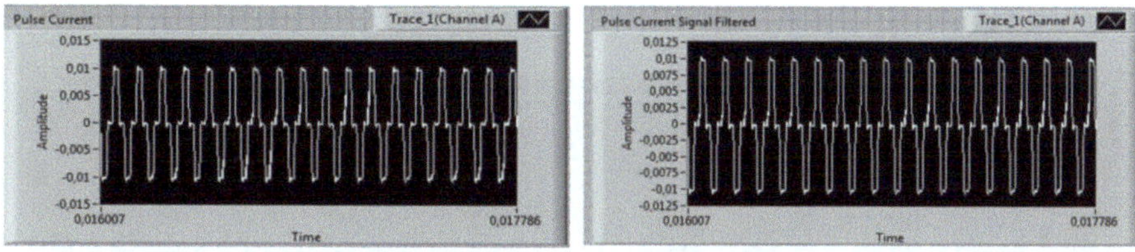

Fig. 8.14 Programmable gain amplifier output with pulse current signal (*left*) and FIR output (*right*)

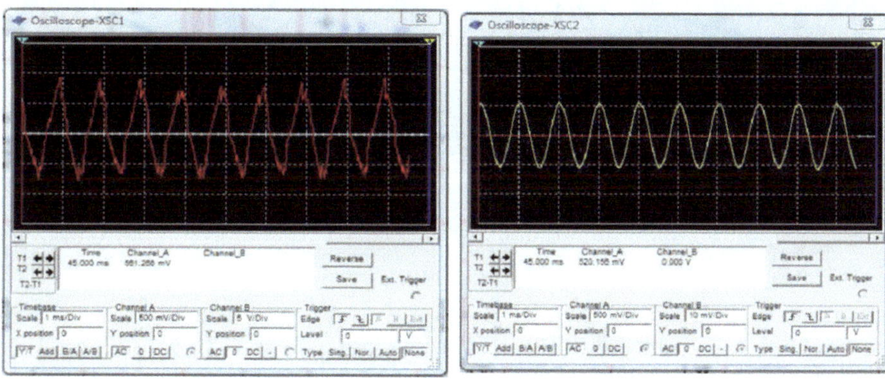

Fig. 8.15 Input sine signal + noise (*left*) and filtered sine signal using elliptic filter (*right*)

a PMT, was used to design all blocks situated downstream with respect to the PMT. The results of the design, owing to simulations, are illustrated in Figs. 8.14, 8.15, 8.16, 8.17 and 8.18. They are clear and illustrate different signal features in either a qualitative or quantitative way. The laser beam is processed and filtered as a result of a variation in gain using a PGA.

8.4 Practical Example

As indicated at the beginning of this chapter, it is possible to detect noise in a bidimensional representation using images. We propose a denoising process of interfence fringes produced by a laser in a thermovelocimetric chamber for soot detection and measurements (Jenkins et al. 2002). A laser He-Ne, it is used in a qualitative manner to determine soot topography during diffusion flame–vortex interactions. The laser is similar to laser-induced incandescence (LII), which is an optical technique (Cetegen and Basu 2006) that has been utilized over the past decade for the measurement of soot topography and soot volume fractions instead of line-of-sight laser extinction and pointwise light scattering techniques. Scattering and extinction measurements are more suited for steady flames and become

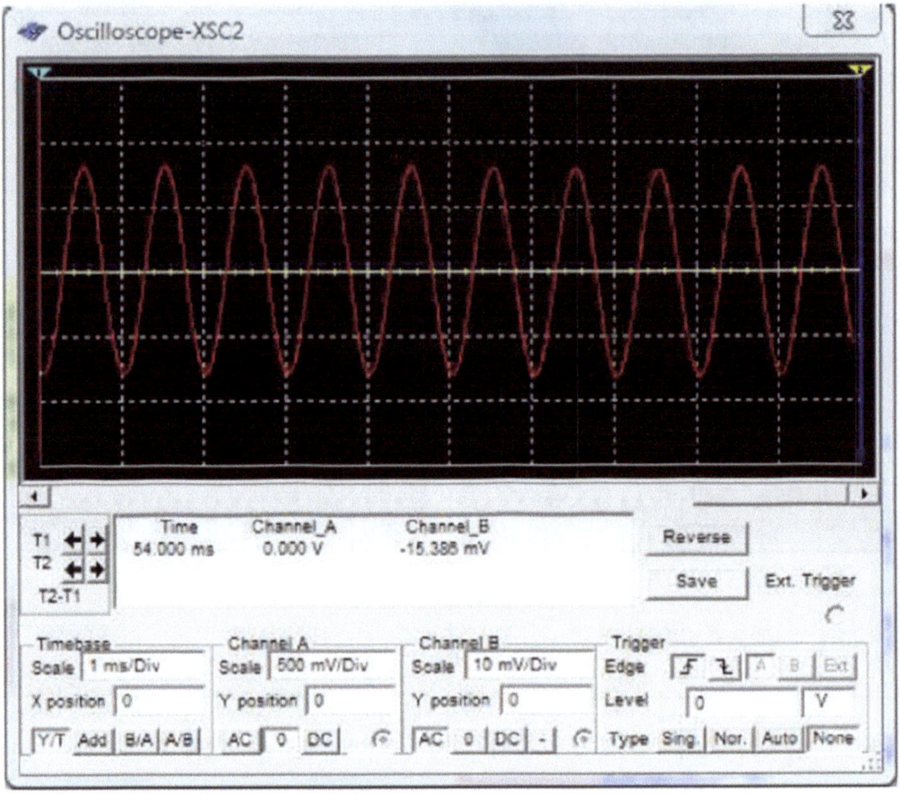

Fig. 8.16 Programmable gain amplifier output

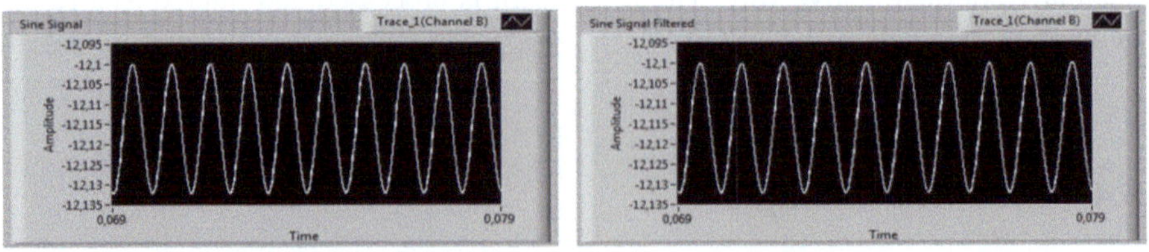

Fig. 8.17 Programmable gain amplifier output with sine current signal (*left*) and FIR output with sine current signal (*right*)

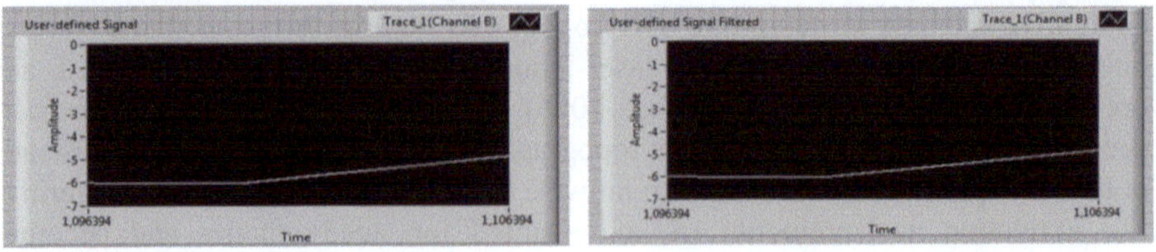

Fig. 8.18 Programmable gain amplifier output with sine current signal (*left*) and FIR output with sine current signal (*right*)

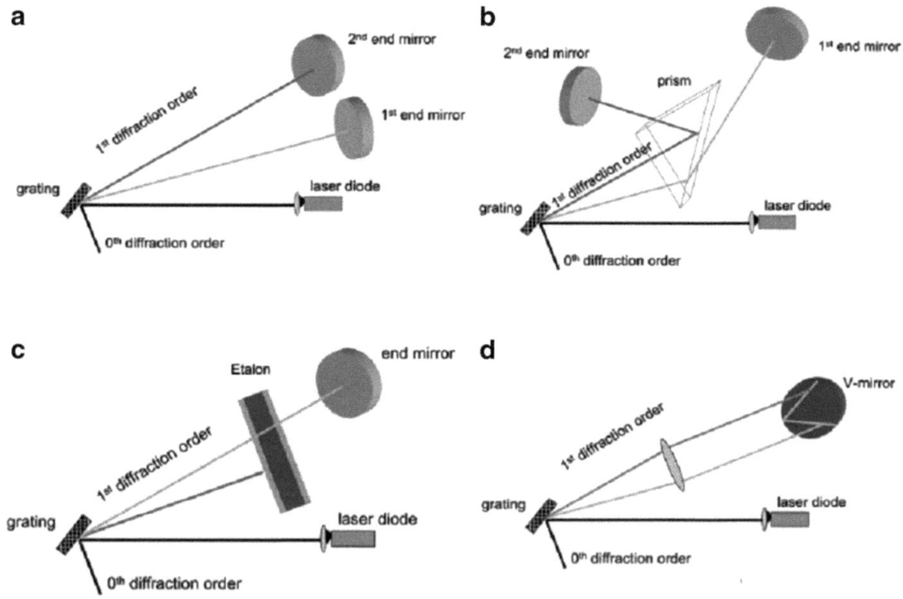

Fig. 8.19 Two-color laser-induced concepts: Double-Littman setup (**a**), Double-Littman prism setup (**b**), Double-Littman etalon setup (**c**), FTECAL setup (**d**) (Friedrich 2008)

more difficult to implement with unsteady flames such as flame–vortex interactions. The two-color laser-induced technique was mostly used for this experiment, along with an absorption measurement (Ciro et al. 2006) as a validation. The two-color laser-induced technique is shown in Fig. 8.19. The experimental setup for detecting soot concentration is described in Fig. 8.20. The fuel is pumped in a velocimetric chamber. Its pressure is regulated by the pump and detected by a pressure sensor. The FPGA serves as a process controller for all operations to be performed in the experiment and is connected to a computer. The laser beam changes with the beam expander and a shutter is placed along its path. Figure 8.21 illustrates the other components of the architectures by means of the velocimetric chamber.

All images during combustion in the chamber are acquired by means of a camera that monitors two beams: the first one is related to a beam that does not pass through the flame that is represented by I_0 as a laser reference according to (8.10). The second one is conditioned by the soot that is produced in the velocimetric chamber. The results of testing are illustrated in Fig. 8.21, where we can see the laser reference image with no soot, a pattern of calibration that is very important for knowing the real amount of soot that must be detected during measurements. As is shown, the soot concentration is captured thanks to a dedicated image that indicates the distribution of soot on the image by a dark color. The remaining area of the image is occupied by fringes of interference that can be reduced according to the image of the right wing of Fig. 8.22.

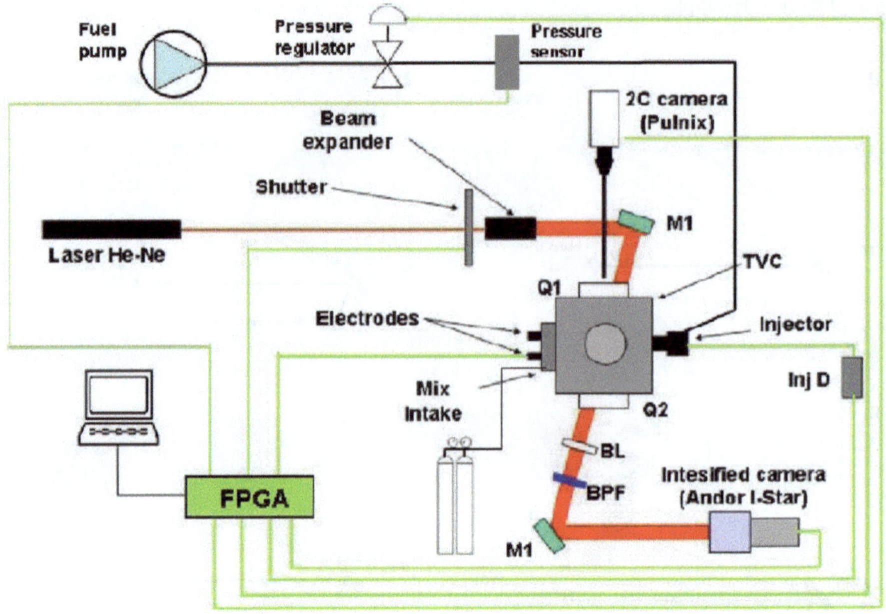

Fig. 8.20 Experimental setup for analyzing combustion gas using laser for a comparison between absorption and two-color laser-induced techniques

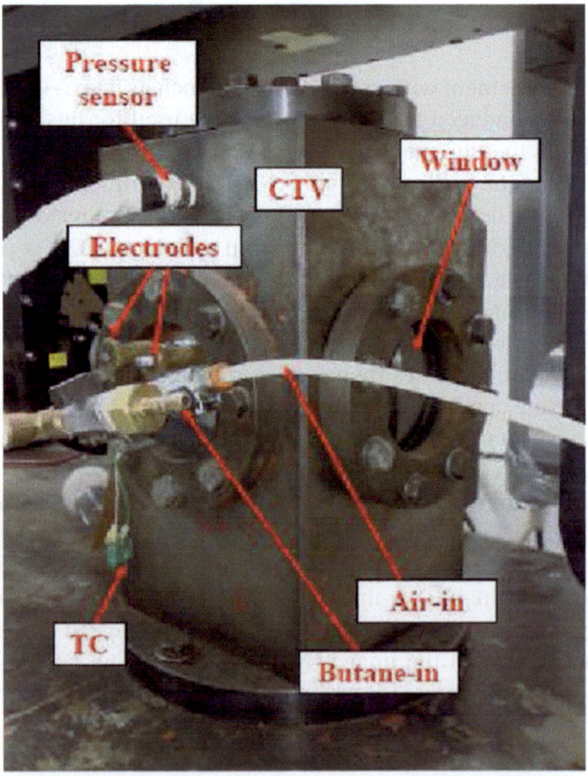

Fig. 8.21 Thermovelocimetric chamber for experimental laser measurements

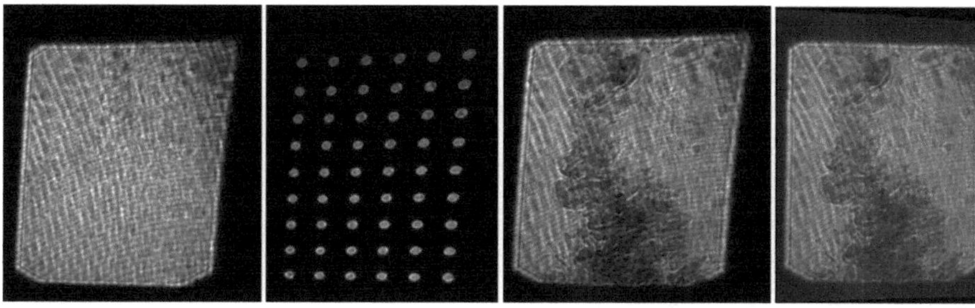

Fig. 8.22 Images representing laser reference I_0 (*far left*), perspective calibration pattern (*second* from *left*), soot absorption with interference fringes (*third* from *left*), and denoised image of soot absorption (*far right*)

As in (7.10)

$$I = I_0 e^{-k.c.d}, \tag{8.12}$$

where

I = radiation intensity
k = proportionally factor
c = concentration
d = layer thickness

$$I = I_0 e^{-\alpha.l}, \tag{8.13}$$

where α is the absorption coefficient and can be expressed as

$$\alpha = \frac{\pi^2 E(m) C_s D_p^3}{\lambda}, \tag{8.14}$$

in which $E(m)$ is the complex refraction index, C_s is the soot concentration, D_p is the mean soot diameter, and λ is the wavelength. Figure 8.23 shows the soot concentration for two series of repeated measurements, allowing for a comparison between the absorption technique and the two-color laser-induced one. Since data delivered by a two-color laser-induced technique cannot be considered as direct measurements because they are calculated using specific formulae following laser acquisitions, they are affected by inaccuracy so a comparison is necessary with another technique like absorption. However, as indicated earlier, repeatability is necessary for a two-color laser-induced technique.

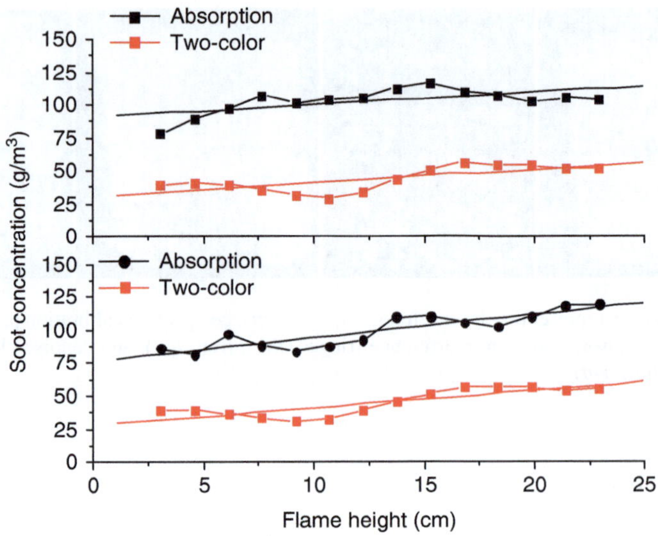

Fig. 8.23 Soot concentration versus flame height using absorption and two-color laser-induced techniques

Acknowledgments Special thanks go to A. De Risi, D. Laforgia, A. Ficarella,. M. Potenza, F. Naccarato, and G. Brida for their time and for putting themselves at our disposal in discussions on some of the issues discussed in this chapter.

References

Cetegen, B.M., Basu, S.: Soot topography in a planar diffusion flame wrapped by a line vortex. Combust. Flame **146**, 687–697 (2006)

Ciro, W.D., Eddings, E.G., Sarofim, A.F.: Experimental and numerical investigation of transient soot buildup on a cylindrical container immersed in a jet fuel pool fire. Combust. Sci. Technol. **178**, 2199–2218 (2006)

Friedrich, C.S.: New two-color laser concepts for THz generation. IEEE J. Sel. Top Quantum Electron. **14**(2), 270–276 (2008)

Haus, H.A., et al.: Representation of noise in linear two ports. Proc. IRE **48**, 69–74 (1960)

http://www.ecvv.com/product/3197740.html

http://www.hamamatsu.com

http://www.ni.com/labview/

http://www.ni.com/multisim/

Jenkins, T.P., Bartholomew, J.L., DeBarber, P.A., Yang, P., Seitzman, J.M., Howard, R.P.: Laser induced incandescence for soot concentration measurements in turbine engine exhausts. In: 40th Aerospace Sciences Meeting & Exhibit, AIAA 2002–0828, Reno, 14–17 Jan 2002

Keithly.: Low Level Measurements, 5th edn, pp. 4.13–4.15

Martin, R.: Noise power spectral density estimation based on optimal smoothing and minimum statistics. IEEE Trans. Speech Audio Process. **9**(5), 504–512 (2001)

Van der Ziel, A.: Noise: Source, Characterization, Measurement. Prentice-hall, Englewood Cliffs (1970)

Index

A. Lay-Ekuakille, *Optical Waveguiding and Applied Photonics*, Lecture Notes in Nanoscale Science and Technology 10, DOI 10.1007/978-1-4614-5959-0,

MIX
Papier aus verantwortungsvollen Quellen
Paper from responsible sources
FSC® C105338

If you have any concerns about our products,
you can contact us on
ProductSafety@springernature.com

In case Publisher is established outside the EU,
the EU authorized representative is:
Springer Nature Customer Service Center GmbH
Europaplatz 3, 69115 Heidelberg, Germany

Printed by Libri Plureos GmbH
in Hamburg, Germany